ASSEMBLING FUTURES

TRANSDISCIPLINARY THEOLOGICAL COLLOQUIA

Theology has hovered for two millennia between scriptural metaphor and philosophical thinking; it takes flesh in its symbolic, communal, and ethical practices. With the gift of this history and the spirit of its unrealized potential, the Transdisciplinary Theological Colloquia intensify movement between and beyond the fields of religion. A multivocal discourse of theology takes place in the interstices, at once self-deconstructive in its pluralism and constructive in its affirmations.

Hosted regularly by Drew University's Theological School, the colloquia provide a matrix for such conversations, while Fordham University Press serves as the midwife for their publication. Committed to the slow transformation of religio-cultural symbolism, the colloquia continue Drew's long history of engaging historical, biblical, and philosophical hermeneutics, practices of social justice, and experiments in theopoetics.

Catherine Keller, *Director*

ASSEMBLING FUTURES
Economy, Ecology, Democracy, and Religion

EDITED BY JENNIFER QUIGLEY
AND CATHERINE KELLER

FORDHAM UNIVERSITY PRESS ❦ NEW YORK ❦ 2024

Copyright © 2024 Fordham University Press
Cover art: Yohana A. Junker, *DesSoterramento III*, 2019.
Mixed media on paper (8.5 in x 11 in) © Yohana Junker

DesSoterramento III

How do we shelter what is precious, yield what is valuable, dispose of what is harmful?* How do we (re)assemble our desires for a livable future while disposing of the waste, violence, and trauma we have caused precious biomes, beings, memories, and histories? *DesSoterramento I–IV* are invitations to dig deep into the terrains of our ontologies, memories, and choices to expose how we have failed to seek and protect life in "this here place." It discloses the possibilities for rerooting and relating to our disciplines, relations, and the earth as dynamic, ever in flux, and emergent. These works ask us to mobilize reciprocity and stimulate care for what is visceral, what is buried, what is hidden. Such excavation will, perhaps, reveal the unseen, dig up the unknown, and point to what has been ignored. They whisper: embrace more people, listen to your kin, sense the mystery. Look closer at what is growing and what is dying. Put bodies and memories in motion. Activate perspectives from down under.

* The artwork and reflection are based on the writings of Robert Macfarlane, *Underland* (2019); Dian Million, "We Are the Land, and the Land Is Us: Indigenous Land, Lives and Embodied Ecologies in the Twenty-first Century," in *Racial Ecologies* (2018); and Raluca Voinea, "Counter-Landscapes: Where Have All the Ruins Gone," in *Debrisphere: Landscape as an Extension of Military Imagination* (2018).

All rights reserved. No part of this publication may be reproduced, stored in a retrieval system, or transmitted in any form or by any means—electronic, mechanical, photocopy, recording, or any other—except for brief quotations in printed reviews, without the prior permission of the publisher.

Fordham University Press has no responsibility for the persistence or accuracy of URLs for external or third-party Internet websites referred to in this publication and does not guarantee that any content on such websites is, or will remain, accurate or appropriate.

Fordham University Press also publishes its books in a variety of electronic formats. Some content that appears in print may not be available in electronic books. Visit us online at www.fordhampress.com.

Library of Congress Cataloging-in-Publication Data available online at https://catalog.loc.gov.

Printed in the United States of America
26 25 24   5 4 3 2 1
First edition

CONTENTS

Introduction | *Jennifer Quigley and Catherine Keller*   1

Our Place on Earth: Territory, Property, and the Sources of Human Entitlement | *Paulina Ochoa Espejo*   7

Democratic Socialism in the USA: History, Politics, Religion, and Theory | *Gary Dorrien*   26

Regifting the Divine Economy: Transitioning Petroleum-Based Energy Regimes | *Marion Grau*   46

The Immanence and Transcendence of Christianity, Capitalism, and Economic Democracy: Alternatives to Ecological Devastation | *Joerg Rieger*   64

Sacred Obligations: On the Theopolitics of Debt and Sovereignty | *Devin Singh*   83

Curating Futures: The Curatorial as a Theological Concept | *Daniel A. Siedell*   106

The Costs of Citizenship: *Politeuma* in the Letter to the Philippians | *Jennifer Quigley*   127

Ambiguous, Amorous, Agonistic, Not Able: An Alternative to Adamant, Apathetic, Antagonistic, Able Society | *Eunchul Jung*   142

What Does Evolutionary Biology Tell Us about Relationality as a Basis for Economics and Politics? | *Marcia Pally*   162

In Whose Interest? Matthew 25:14–30 as a Theo-Economic
Parable Hard at Work | *Hilary McKane* — 183

Creeps of the Apocalypse: Climate, Capital,
Democracy | *Catherine Keller* — 201

List of Contributors — 219

Index — 223

ASSEMBLING FUTURES

◆ Introduction

JENNIFER QUIGLEY AND CATHERINE KELLER

Amidst the fitful dramas of democracy, does it make sense to assemble such a volume as this? The speed of political struggle now persistently transcends the rhythm of book publication. Our timely examples of the interplay of economics and democracy with, for instance, the polarized swings of U.S. politics, including the judicial take-down of Roe v. Wade, or with global responses to Russia's invasion of Ukraine, or with the predictably accelerating effects of climate change, cannot hope to be quite up to date. Yet even though there is no keeping up with the present actualizations, our transdisciplinary theological conversation is motivated by the concrete urgency of a now-time whose future horizon looms ever close. The facts and faces of our economics, ecology, and democracy may shift significantly, but the systemic tendencies and the material dangers persist all too repetitively. And so the claim upon the reader's time requires that, despite the lag time of this non-journalistic genre of writing, we not only address the matter of *possible* futures; it requires that we understand this writing to be taking part, by way of the oddly dispersed democracy of reading, in the assembling of futures.

Assemblage, an ancient political trope of egalitarian gathering and deliberation, now carries a conceptual sense of dynamic and systemic complexity. Deleuze and Guattari theorized assemblage as the self-organization of multiply layered, coded, and shifting systems. In their heterogeneity these gathering manifolds form "constellations" of myriad elements, organized in material relations of language and power. The liberating potential of an assemblage depends upon awareness of its own dynamic of multiplicity of multiplicities. So, for instance, if enough of us do not know that neoliberal capitalism in its resilience greenwashes its image with shifts far too minor to slow global warming; or that its neocolonial economy feeds upon the legacies of colonial politics; that white supremacism bleeds democracy of self-corrective energies—then change serves a disturbingly fixed future. And white conservative Christianity can thus strengthen the worldly status quo, in resignation or in triumph, with the promise of the one big change—to a changeless otherworldly future.

When we began organizing a Transdisciplinary Theological Colloquium entitled "Assembling Futures: Economy, Ecology, Democracy," planned for March of 2020, our conversations took place in a different future than we could have possibly imagined. The gathering of a group of scholars and students from around the globe was quickly scrapped in the emergence of the global COVID-19 pandemic, at the time expected to cause only brief disruptions to calendars and lives. We were able to reassemble virtually, however, in fall 2020, to pursue the delayed but ever more timely conversations around these themes. We now assemble these essays in the dynamic and systemic complexities of possible futures that seem ever closer to the present. We live our lives innocent of the future, to paraphrase biblical scholar Paula Fredricksen's description of the Apostle Paul, but we write, edit, and deliberate from across spatial and temporal distance to contribute to an assembling future with the ethical urgency that the now requires.

These essays, as with all the Transdisciplinary Theological Colloquium volumes, reflect scholarly conversations among historians, political scientists, theologians, biblical studies scholars, and scholars of religion that transgress disciplinary boundaries to consider urgent matters expressive of the values, practices, and questions that shape human existence. The scholars participating in *Assembling Futures: Economy, Ecology, Democracy, and Religion* recognize urgent imbrications of the global economy, multinational politics, and the materiality of ecological entanglements in assembling still possible futures for the earth. In the firm conviction that the urgency of the future requires not siloing our imaginations, we have assembled these essays without dividing them along traditional disciplines or themes. Precisely in their diversity of disciplinary starting points and ethical styles, the essays that follow enact their intersectional force field all the more vibrantly.

In "Our Place on Earth: Territory, Property, and the Sources of Human Entitlement," Paulina Ochoa Espejo reframes the debate in political theory about territorial rights by centering indigenous understandings of subjects. She offers an enlivening reflection on the notion of territorial rights reconstrued in terms of the *pueblos* as subject. Territory reveals again its terra. Rather than a binary of private (and individual) versus state's rights, which have so often mismanaged resources such as water, Ochoa Espejo's earthy reframing of rights would be inspired by indigeneities of people, peoples, understood as entangled with plants, animals, territories, waters. Using the example of rural Mexican *pueblos* as just such an entanglement, she concludes that the symbol of a harmonious water cycle within complex ecosystems reflects how territorial rights belong to grounded and entangled communities.

Gary Dorrien demonstrates the complex but enduring histories of U.S. and European democratic socialism, including Christian socialism. In "Democratic Socialism in the USA: History, Politics, Religion, and Theory," his analysis of the tension on the left between redistribution and recognition—that is, between political economy and the cultural focus on race, gender, and sexual identities—lands needed public wisdom. With a sweeping history ranging from radical democratic movements predating European socialism in the early nineteenth century to analysis of the burgeoning popularity of the Democratic Socialists of America among Millennial and Gen Z activists and voters, Dorrien imagines possible futures for democratic socialism and for its influence upon American democracy and its economy.

In "Regifting the Divine Economy: Transitioning Petroleum-Based Energy Regimes," Marion Grau critiques the anointment of power with oil—petroleum's sacramentality in neo-liberal, economic, and consumptive structures. She examines the powerful grip of petrol oligarchies and their orthodoxies by reconsidering gift economies. She leaps to the heart of the conceptual dilemma with Derrida's *pharmakon* of the impossible gift, performing a deconstruction of any orthodoxy of the one-way grace, even in the Derridean one-way gift. Building on concepts emerging from indigenous gift economies such as potlach traditions, Grau reimagines the category of gift as *Eu-Charis* (the good gift) in ways that indeed are freely given but oblige to responsible reciprocity if the gift is to be honored and life to be valued.

Joerg Rieger considers how both the Gospel of Prosperity and Christian capitalist complicity in ecological devastation belie the routine liberal/progressive claim that the answer lies in focus on the immanent and material. Neoliberal capitalism shares much of that latter focus. In "The Immanence and Transcendence of Christianity, Capitalism, and Economic Democracy: Alternatives to Ecological Devastation," he demonstrates the importance of reconstructing transcendence along with immanence. Situating his work within the new materialism emergent in the study of religion, Rieger urges theorists to keep sight of the dynamics of production and the exploitation of labor, despite shifts toward the finance-dominated neoliberalism that Kathryn Tanner names in *Christianity and the New Spirit of Capitalism*. He makes key the material relation of the exploitation of labor to exploitation of the earth. Rieger thus models how to revisit fundamental categories such as immanence and transcendence while investigating power and agency in economic and ecological democracy and democracy's embodiment in political, cultural, and religious forms.

In "Sacred Obligations: On the Theopolitics of Debt and Sovereignty," Devin Singh considers the role that understandings of the sacred and the religious have frequently played both in affirming and in challenging connections between debt and sovereignty. Singh notes particular historical moments, as in those of Ancient Near Eastern kings or of Greek philosophers, that highlight the fact that sovereignty thrives on the skillful administration of debt; "inasmuch as divine kingship has left its imprint on medieval and even modern conceptions of sovereignty and governance," he concludes, "the divine economy and its heavenly creditor have also left traces that contribute to debt's ongoing centrality and influence in Western societies today."

Daniel A. Siedell deploys the visual arts as a medium for theological and imaginative responses to the many crises, catastrophes, and apocalypses related to climate, politics, and the economy. "Curating Futures: The Curatorial as a Theological Concept" lifts up the role and image of the curator as opposed to that of the artist as sovereign creator ex nihilo. Siedell argues that the very act of assembling futures requires curation rather than creation. Using the complicated intersections of corporate pressures, activist artists, and curatorial choices at the Museum of Modern Art (MoMA) as a case study, Siedell contends that curators and theologians have much to learn from one another about assembling temporary relationships among the irreconcilable multiplicities of ecology, economics, and political apocalypses.

In "The Costs of Citizenship: *Politeuma* in the Letter to the Philippians," Jennifer Quigley considers the singular occurrence of citizenship in the letters cowritten by the Apostle Paul as a lens through which to consider the complex layers of citizenship and its relationship to politics, ecology, and democracy. She reads the Letter to the Philippians, where Paul claims the Philippians have "citizenship in heaven," with attention to local archaeological detail and the material realities of a small, impoverished Christ-community living disenfranchised in a marshy backwater of the Roman empire. Quigley argues that while the invocation of a heavenly *politeuma* and a divine savior figure might have been limited by the contextual constraints of living under empire, the Philippians and Paul were beginning to imagine and assemble a potential future economic theology in which financial/legal exploitation is replaced with divine abundance.

Eunchul Jung's "Ambiguous, Amorous, Agonistic, Not Able: An Alternative to Adamant, Apathetic, Antagonistic, Able Society" explores the endemic persistence of tribalism. Building upon the work of Robert Corrington and Catherine Keller, Jung explores the surprising correlations between tribalism and global capitalism. He finds his assonant alternatives in two counter-strands of ethics, those of Byeong Chul Han and Jack Halberstam. He imagines a liberative future emerging not from an already existing system, but through the queer art of failure and *being able not to be able*.

In "What Does Evolutionary Biology Tell Us about Relationality as a Basis for Economics and Politics?", Marcia Pally describes an ontology of relationality as a baseline for considering what might contribute to our thriving in our ecological setting. She compares relationality as a key concept in Christian conceptualizations of the Trinity, in Jewish understandings of covenant, and in biological and psychological descriptions of humans as a hypercooperative species. Interweaving insights from philosophy and theology with recent work in evolutionary biology and developmental psychology, Pally concludes that more explicitly relational economics, politics, and approaches to ecological crisis are not only needed, but are expressive of what characterizes the "baseline" human condition.

Hilary McKane looks at the "parable of the talents" to examine how financial systems informed both the creation of the parable and its attendant theological implications. But the parable then functioned historically to support those same systems. "In Whose Interest? Matthew 25:14–30 as a Theo-Economic Parable Hard at Work" examines the interpretive afterlives of this parable at key moments of theological, political, and economic transition, with treatments of ancient, medieval, and reformation interpretations. McKane concludes, in conversation with Kathryn Tanner's "Protestant anti-work ethic," that "the parable can be read as a cautionary tale against the unbounded pursuit of monetary interest precisely because it challenges us to confront the uncomfortable connections between faith in divine sovereignty and faith in the money market."

In "Creeps of the Apocalypse," Catherine Keller considers the strange temporal interplay of climate, capital, and democracy. From a daily perspective the threats of global warming advance still too slowly to stir the needed speed of response; yet from a geological perspective, indeed even that of a single century, the speed of the changes threatens to gallop apocalyptically past any solutions. On one side of our Anthropocene the choppy temporality of political headlines and electoral deadlines defers the last chances of meaningful action; on the other side the commodified time of neoliberal capitalism covers its carbon emissions with the smooth continuum of assured progress. How can a theological transdisciplinarity break through this triple creep of ecology, economics, and politics for the sake of a still possible polyrhythm of shared planetary life?

You will find your own transdisciplinary connections among these essays, perhaps considering the alternative epistemologies modeled by pueblos and by heavenly citizenship; or examining day labor in a Matthean parable alongside contemporary exploitation of labor and the earth; or contemplating what is unveiled with a critical look at petro-theology amid an apocalyptic Anthropocene. Your participation in the oddly dispersed democracy of reading and the connections you make collaborate with us in the assembling of futures.

❧ Our Place on Earth: Territory, Property, and the Sources of Human Entitlement

PAULINA OCHOA ESPEJO

"Day Zero"—that is how citizens of Cape Town, South Africa, referred to the day in which the dams would go below 13 percent—that is, the day the city would run out of water. During the worst days of the drought, people would go without washing, and it all came close to the moment in which they would have to begin rationing their drinking.[1] Only a year later, in Chennai, India, there was another Day Zero: the reservoirs were officially declared dry. People lost their jobs because they had to stay home, waiting on the sidewalk for the trucks bringing water from afar. People fled their villages and animals died because of the bad urban planning that destroyed the wetlands around the city.[2] These two cases are spectacular, but sadly, not exceptional. Days Zero have happened or are expected to occur in many other parts of the world. Millions throughout the world lack steady access to clean water, and those who have taps in their homes often depend on unstable or unsafe sources. Soon we will have to make important political decisions regarding water—but who will make them? Should the very states who have made disastrous decisions regarding natural resource exploitation, urban planning, distribution, and conservation decide on the future of these resources? Do property owners have special rights over the water under their feet, and should they be allowed to sell it in bottles? And if we are suspicious about state property and private property of water; if we believe that all humans share the water in common—who will make political decisions regarding communal distribution and conservation? How do we represent the voice and the interests of all humanity? These questions about who is entitled to make decisions over the land and its natural resources, including water, boil down to a more precise question: who has territorial rights? As the examples of Cape Town and Chennai make clear, the question will become more important as the climate continues to change.

Territorial rights are now hotly debated in political theory. But the discussion takes place against a wide consensus on who are the *kind* of subjects eligible for rights. According to the dominant views, either individuals or states can have rights over land and water. They have this right because human beings (and the political collectives meant to enforce individual rights) are entitled to the Earth. That is, most theorists hold that humans are inherently deserving of the privilege of mastering and controlling the land and its resources. There are debates over which particular humans have special rights over which particular lands, and debates over which specific individuals and groups are connected to some parts of the Earth. But territorial rights theorists widely agree that the ultimate source of entitlement is the inherent value of human individuals, which rests in a metaphysical or religious view of human agency, reason, and/or the special place that humans in the world hold vis-à-vis other creatures and things.

However, there are other views about the sources of human value, and thus, there could also be other conceptions of territorial rights. Besides the political theologies that undergird the main position in current debates, there are other political theologies that are not anthropocentric or centered on the idea of mastery and control. One such political theology finds the sources of human entitlement and the idea of territorial rights in the entanglement of people, plants, animals, and things. This, in my view, can be symbolized by the harmonious water cycle within ecosystems. Following this view, we could hold that territorial rights do not belong to either states or individuals, but instead to grounded (or entangled) communities.

This chapter explores this alternative. It distinguishes entangled sources of territorial rights from the more common views grounded in Sovereignty or Natural Law; and it shows how these entangled groundings for territorial rights have been used in the past by many different groups of people connected to places. One such use comes up in a concept that is recognizable to indigenous communities across Spanish America: the concept of *pueblo*. "Pueblo" means both "town" and "people" in Spanish. A *pueblo*, in the way I propose we interpret it, can be the subject of territorial rights because it exemplifies how a human community is entangled with plants, animals, and things—all of which are deserving of respect. This is a type of grounded community where people recognize their located entanglements and the obligations to the place where they are, the air they breathe, and the water of which they themselves are made. These communities could be the locus of important decisions regarding land and water soon when dealing with the planetary climate emergency.

In the rest of the chapter, I explain the theoretical and political problem of territorial rights in the time of climate change. I show the theological underpinnings of the main philosophical views that frame territorial rights as private property. I describe how a different political theology—a view inspired by Keller's *Political Theology of the Earth*[3]—could provide other grounds for territorial rights and for motivating coordinated action. At the end, instead of a conclusion, I make a case for considering *pueblos* as a subject of these territorial rights. Decisions about water and place should also be made by those grounded communities who see water as the messenger letting us know of our proper place on Earth.

THE POLITICS OF WATER AND THE QUESTION OF TERRITORIAL RIGHTS

"Day Zero" symbolizes water scarcity, but such scarcity is not the only concern. Scarcity is likely to give rise to political conflict. For example: in May 2014, a violent confrontation broke out between, on the one hand, residents of San Bartolo Ameyalco, a mountain village that has been engulfed by the urban growth of Mexico City, and, on the other, the Mexico City police. The police were protecting city workers in charge of digging trenches for the development of a hydraulic system. The system was intended to pump water from an aquifer that had traditionally been managed by the authorities of the village. The city asserted rights to the water, which, officials claimed, was intended for parts of the city without access to the water grid. However, the residents of San Bartolo claimed that the city government intended to use the water for new development of luxury condominiums and high-rises in neighborhoods far removed from San Bartolo. The media stressed the economic differences between the village and the rich neighborhoods for which the villagers claimed the water was destined; but after the confrontation, it became clear that there had been many tensions within San Bartolo, too. A faction of San Bartolo residents benefited from the sale of water from the local spring, while poor newcomers did not have access, either to the aquifer or to the water grid.

These events did not get much media uptake outside Mexico City,[4] but they seem increasingly familiar. Across the globe, we find many other instances in which a complex set of political interests comes to the surface when local rural residents resist the state's attempts to exploit natural resources for the benefit of urban folk. Cities need water and energy, and the extraction and distribution of these resources often intensify problems of inequality, social marginalization, and other forms of injustice. The violent confrontation in San Bartolo is only one example of a trend that will intensify in the future. As the world's population grows and its climate changes, the question of who should be entitled to control access to land and natural resources becomes even more urgent.

Moreover, control of natural resources is tightly woven with other political questions. Those who decide on how natural resources will be allocated are those who ultimately decide on the questions of territorial rights. Decisions regarding natural resources go hand in hand with other decisions about border control and territorial jurisdiction. Who owns the land? Who has a right to exclude people from the jurisdiction? What are the conditions in which people could be excluded? Are those decisions private or public? Should they be mediated by the state? Can they be resolved without the rule of law? Because these questions about resources are so tightly connected to matters of jurisdiction, it is impossible to disentangle the question of who should have resource control from other touchy political issues such as the strategy of national economic development, immigration and asylum, border control, and the rights of indigenous peoples.

These political ramifications beg for analysis, and the problems they engender require new creative solutions. Political thinkers, ideologists, and critics need to think about these problems again, in terms of the ground of territorial rights, because the ideological tools that we have at our disposal are not suited to the new crises. This is clear when we think that climate change must be addressed from a global perspective and water cycles cannot be kept within one country. Yet we lack the vocabulary and the motivating narratives to lead us out of the problem. The politics of planetary change are caught up in the politics of water, which remains within the limits of the existing patterns of thought in traditional theories of territorial rights.

When political problems related to resource use are discussed in the public sphere, there is a widespread tendency to connect them to questions of territorial rights and, in turn, to narratives of individual control and the exclusion of others. That is, when it comes to resource scarcity, the standard reaction is xenophobia. The clearest example is that for many people the reflex response to environmental threats is to close borders to outsiders. Territorial control becomes a xenophobic policy of border control meant to preserve resources for those inside the territory. This reaction presupposes the view that resources belong to those who *own* the land. The politics of territorial rights and natural resources goes hand in hand with the politics of ownership and property.

TERRITORIAL RIGHTS AS PROPERTY

Questions related to natural resources (particularly water) and their political ramifications are typically studied using theories of property as theoretical ground. In the last twenty years, political philosophers have developed new theories of territory that speak to thorny political questions. These include theories that ground their proposals on individual private property,[5] on the historical attachment of peoples to territory and to the land,[6] and on the obligations that individuals owe to each other and that we can only discharge through the state.[7] These theories illuminate many existing intuitions about jurisdictional rights, and they also address complex issues regarding resource rights and justice in the era of climate change.[8] For example, some hold that we must share resources because humans have original common ownership of the Earth,[9] while others believe that sovereign peoples or states have exclusive rights to their land and their natural resources.[10] Yet, even though they disagree over crucial matters, these theories all model territorial rights on the individual right to private property.[11]

The individual right to private property is the foundational right on which collective rights (such as the territorial rights of states) are modeled. Private property is a system of rules that allocates things or "land to particular individuals to use and manage as they please, to the exclusion of others."[12] Each individual who has title over a particular thing or tract of land has a right to use it or dispose of it, regardless of the impact that her decision may have on others. Not all theories of property rights assume that individuals have complete mastery over their property. In fact, the dominant legal and philosophical doctrines recognize that private property rights are often curtailed by the state (there are rights of way and legal limits on construction and use), and they describe property as a "bundle of rights" rather than as a sole title.[13] However, they all assume that property is a relation that gives an individual or a group the power to dispose of an object. They also presuppose that this object can be inherited: an owner's heirs have a right to use and exclude from their forefathers' property.

Many scholars of territorial rights have emphasized the differences between private property and collective ownership. However, they still envision territorial rights (which are collective) on the model of private property. The main difference is that property is individual and assumes the existence of other individuals who recognize and respect this right, whereas territorial rights are collective, and they presuppose the existence of a political authority that asserts legal control (jurisdiction) over a certain area.[14] However, even if we make a clear distinction between private property and political jurisdiction, the states' territorial rights are still modeled on individual private property: both have the same structure. In both cases, collective ownership is imagined as a bundle of rights that a subject (individual or nation) has over a detached object (land or natural resources). Territory is understood as the nation's property; it is exclusionary, and it is inheritable.

The fact that state collective ownership is similar in structure to private property is important because nowadays, most people think that only states have title to territory and only states have the legal right to jurisdiction, to control natural resources, and to control the borders of a given country. So, the property view permeates most theories of territorial rights. It is true that in the last twenty years, many political philosophers have challenged the view that only states have a right to territory. This challenge comes from the fear that the state is too closely allied with the interests of capitalism. The state can support the interests of commercial corporations against the interests and rights of collectives other than the state, such as peoples, nations, tribes, or cultural groups. Yet even these challenges still hold that territorial rights must be based on *title* to territory. For example, when indigenous groups such as the Standing Rock Sioux Tribe protest economic interest over their land in association with the state, they often do so on the grounds that it is the tribe who owns the land. Yet, the idea of title is a notion taken from the law and morality of property—it is a claim to dominion. Hence, both the established view and its challengers tend to see territorial rights in terms of the most intuitive property right: the individual right to private property. They all tend to see the right to territory as in fact a bundle of rights or incidents that allow for dominion over land. However, as recent climate changes remind us, the environment is not an object separated from individuals that can be dominated and controlled; nor is it a fungible resource that can be exchanged in the market.

This is a failure of our theoretical imagination. For many people across the globe, the political appeal of protest movements against capital and the state, such as the protests of the Dakota Access Pipeline, is not solely grounded on the intuition that the Sioux (or other indigenous people) are the rightful owners of the land and that as rightful owners they can do whatever they want with it. This is clear from the fact that other indigenous groups that seek to industrially exploit oil and other natural resources did not get the same kind of political support. For example, support for the movement did not come as easily to the Native people who wanted to dig for oil on their land in Canada.[15] The intuition of many of those who protest (including many indigenous people) is that they oppose the view that territorial rights entail exclusive jurisdiction and the right to do whatever the owners want with the land they possess. Many reject the idea of dominion and control of just one group, and they reject the idea that this group has the right to dominate, control, and dispose of the land as it sees fit. The intuition of many of those who protested is that humans do not *own* the land, that we all share the world where we live in with other living creatures, that the combination of people, biota, and things has an intrinsic value beyond the value that rulers decide or markets determine.

Envisioning territory by analogy to property is a problem because this model leads to a dead end in a world wrecked by climate change. When we see land and resources as an object independent of its owner or a fungible good in a market, then we are beset by insuperable problems of distribution, or we tend to destroy the very object that we value. For instance: we recognize the value of the carbon sink when dealing with climate change, but it is difficult to justly distribute shares of carbon sinks, particularly when these are seen from a historical perspective.[16] We know all human beings have a moral and a legal right to water,[17] but water cannot simply be evenly distributed among individuals across the globe, because when water is taken out of the natural systems that reproduce it, it ceases to be a valuable resource.

We know we must mitigate the effects of climate change, adapt to them when they come, and deal with the effects we cannot avoid anymore. If so, however, then we must integrate complex environmental processes into our thinking about territory and territorial rights. To do this, we must model territory and territorial rights on something other than property and rely instead on models that take seriously the environment and people's connection to it. But could we think about territorial and resource decisions beyond property?

Many of those who protest in Standing Rock and in other similar battle sites think of natural resources in terms of *commons*. However, common ownership remains an idea tied to ownership—that is, to property. And if property remains at the center, so will the idea of dominion and control. This is clear if we think of common property as that to which we all have equal access. This means that common property, particularly universal common property of natural resources such as water, is likely to be abused not just by one owner, but by all. This is the problem that economists call "the tragedy of the commons." The traditional way of dealing with this problem is to appeal to an external source of governance to prevent abuse by individuals. Thus, the commons end up governed by the market, the state, or the community. But these traditional solutions encounter many difficulties at the global scale; most people believe that state governance is a non-starter, given that a world state is undesirable; a global private market of shares of the Earth does not make sense, given that it is impossible to divide land in equal shares or to distribute evenly the burdens and the responsibilities related to fluid resources such as water and air. (Can we each exclusively own our private share of the atmosphere? Can we have individual shares of ocean or the wind?) And finally, we don't have a worldwide common culture on which to rely to determine a common set of behaviors beyond the threshold of national or international law. The problem of thinking of the Earth in terms of commons then, is that commons are still common *property*. Although many have an important intuition that we don't own the world, it is hard to come up with an account that does not see the land and natural resources as some form of property—individual, national, public, common . . . but always property.

THE POLITICO-THEOLOGICAL ORIGINS OF TERRITORIAL RIGHTS AS PROPERTY

Most of us remain committed to seeing territory on the model of individual private property. This is probably connected to historical arguments that legitimize possession of the land as an individual right to personal property and, ultimately, to politico-theological views associated to either natural law or sovereignty.

The first view—as we know it today—emerged in European Enlightenment thought as a reaction to theological ideas about the origin of law. In Medieval Europe, territory was often imagined as divinely given and inherited as a personal property of kings. Territorial legitimacy was part of divine right—a title of royal families originally given to Adam and passed down through primogeniture across the ages. This idea of territorial legitimacy based in divine right sounded absurd even to seventeenth-century ears (certainly to Locke's, who attacked this view in his *First Treatise of Government*).[18] But in the new enlightenment discourses, territory was still imagined by most philosophers in terms of personal property and associated with each individual's capacity to work on the land and transform it or to show their original title by virtue of continuous occupation. This view was grounded in natural rights and ultimately in natural law. So, although God gave the Earth to all humans in common, as Locke and Grotius believed, the land ended up parceled out and fenced and distributed among different countries. In this natural rights view, the parceling happened because each of us has a claim to one's own body and one's own labor; through labor we can legitimize private appropriation and claim a right to hold private and collective property.[19] This popularized natural-rights account of territory has accompanied many claims to land in the world since the sixteenth century: the idea that the land belongs exclusively to those who toil it or occupy it has been put forth by many groups, from Northern European peasants to American colonists.

Although territorial claims in modernity can differ widely and fit a whole range of political views and concerns, they are mostly tied to this idea of private property and natural individual rights to ownership. Even those who critique private property seldom challenge the core of the source of entitlement: the idea that humans own their own person and that humans alone endow the world with meaning and value. This anthropocentrism also gives humans the right to own and to command nature.

This can be illustrated further through natural rights ideas. As the Lockean argument goes, we are God's creatures, but we own our body and the fruit of our work just as God owns us. God as a divine maker commands, controls, and dominates the world. And given that humans are made in God's image, they too command, control, and dominate. So, according to this view, whether territory belongs to the nation, or to the individual, or to the state, the owner has dominion and the right to control. The property owner remains a miniature version of the king on his throne.

The natural law tradition has persisted within liberal thought in secularized versions of the rights to individual property that entered the political imagination of the West through the work of Spanish scholars Vitoria, Acosta, and Suárez and the protestant natural law theorists Grotius and Locke. This political theology, which sustains the idea of territory as the outcome of individual property, is built on the tradition of natural law, but this is not the only theology available. There are other political theological ideas that have persisted in modernity. A well-known second view undergirds legal positivism and sees territory as state jurisdiction.

This second view envisions territory as the collective property of a people within a state. It differs from the first because it does not imagine property as a natural right of individuals. Instead, it sees property as a conventional arrangement authorized by law. The rules that govern ownership come directly from the lawgiver, who is also responsible to grant titles. Thus, private property is secondary to territory and jurisdiction: the result of the recognized authority of a sovereign that establishes the ground of law in the state. Property law, then, is not natural, it is conventional, and it requires the authoritative commands of the sovereign that enforces contracts among individuals. This view is associated with Thomas Hobbes and other political philosophers who imagined the sovereign as the origin and the center of the state. Here, territory does not emerge when individuals bring their private properties together. Rather, state jurisdiction is a prior requirement for property. Private property depends on individual title to parceled areas, and the state's capacity to subdivide and keep contracts among individual owners is in turn the result of the successful establishment of authority over a territory. Sovereign authority goes hand in hand with political control. Authority depends on the state's capacity to be the sole arbiter and apply its law in each land. The state "owns" the territory because it has exclusive jurisdiction, and it can control the land and exclude others from it.

This second view, moreover, is also the result of a particular account of the divine and its relation to the world. According to the legal scholar's tradition of political theology,[20] this conventional view of property is contained within legal positivism, which is a secularized version of theological voluntarism. According to this way of understanding political theology, the sovereign's authority comes from an analogy (or structural relation) to God's authority—specifically, an analogy to the traditional image of God-as-Judge. The juxtaposition of King-as-Judge on its throne, with God-as-Judge overlooking creation, eventually produced the image of the Sovereign as an all-powerful ruler of the kingdom or the state (as has been extensively argued by Schmitt and Kantorowicz).[21] In this account, positive law is a secularized version of divine law, and the sovereign is a secularized version of a voluntaristic God. Territorial rights, or rights of jurisdiction, presuppose absolute mastery and control, given that they are a secularized version of God as sovereign over the world. Just as God-as-Judge stands on his throne before creation, so does the sovereign on his throne, wielding complete control over his kingdom, the land, and its resources.

The dominant traditions that ground territorial rights (private-property natural law and public-property sovereignty) differ in many ways, but both rely on arguments structurally analogous to theological views of human entitlement. They both see the land and resources—particularly water—in terms of ownership. Nature is to be mastered. These views prefigure how political philosophy conceives of the relations of individuals and the land on which political organizations have jurisdiction. Background political theologies supply narrative patterns that shape territory as property.

A DIFFERENT POLITICAL THEOLOGY: OTHER GROUNDS FOR TERRITORIAL RIGHTS AND MOTIVATING COORDINATED ACTION

Could a different political theology supply an account of the political relation of people and nature that is better suited to connect to our ravaged Earth? The view that I propose here is a political theory of place, inspired by Catherine Keller's *Political Theology of the Earth*,[22] and many of her interlocutors, old and new (including Nicholas Cusanus, Alfred North Whitehead, and Bruno Latour). This political theology is an ecosocial perspective that focuses on "the complexity of the web of asymmetrical forces of which our common life is precariously woven."[23] It takes seriously the possibility of finding a source of normativity in nature. In my account, territorial rights are also grounded in theological sources of human entitlement, but here, they come together with ideas of human and divine entanglement with the materiality of the world. In the politics I want to endorse, the ground of territorial rights is conventional (as it is in the positivist version), but conventions are not the source of power of the state, and law is not a command from a sovereign lawgiver. Instead, the source of rights comes from a common agreement that is ultimately grounded in place, buoyed by nature and its internal entanglements with the divine. This source of norms can be symbolized by the harmonious water cycle within ecosystems.

This different political theology comes at the end of our anthropocentric era. It arises when all humans are forced to re-evaluate human needs and human interests. As we run out of water, we are forced to bring our thought and our belief back to the entanglement of all creatures and things. This does not happen only because we have come to recognize other creatures' intrinsic value—a value that they have and that more of us now acknowledge—but it is also a result of the self-interested realization that unless we take other creatures into consideration, human beings will face incalculable losses. Mastery and control are not a rational solution once we realize that disposing of the Earth as we see fit will inevitably destroy the very resources we need.

However, this realization alone is not sufficient to motivate coordinated action. To do that, the end of anthropocentrism must have new ideas and emotions. One way to find those is to supplement scientific descriptions of the current state of the world with philosophical and theological accounts, particularly with those ways of thinking that had long recognized that the value of the human species does not reside in its exceptionality, but rather in its capacity to comprehend the value of its relations with other creatures. For these ways of thinking, the value of individual action emerges only in its relations to others—other people, other creatures, other things. If there is anything exceptional about humans, it is their capacity to comprehend that they have a role to play in such a system of systems, which makes the Earth function as a harmonious whole. Motivating action along these lines is one of the important political roles that such theology could play today.

Yet, the most important part for motivating coordinated action is not in human understanding, but in human experience—specifically, the experience that within the whole that is Earth, each distinct part (if there are such distinct parts) sustains itself in equality and cooperation with others. Motivation to act toward preserving the whole must be grounded in the feeling that each part may have a say in a "democracy of fellow creatures."[24] This experience can help us cooperate and eventually find common ground to act to preserve ecosystems that produce water, and from this action we may also come to understand territorial rights in the very terms dictated by water without having to conceive of the land as an object that an owner possesses.

When we ask who has territorial rights, we are also asking who will make decisions regarding air and water. Will it be property owners? Will it be the state? Will it be the whole human community? The entangled account that I propose holds that the lakes and the rivers, water itself, may have something to say about these questions. Nature itself may be the source of value that helps people make political decisions. This idea may sound puzzling at first, given that since early modernity, political authorities and philosophies have sidelined any sources of authority that do not clearly spring from human reason. Since early modernity, nature and the divine were moved aside as man became the measure of all things. The value of man and his reason became the ground of political morality and the source of man's entitlement to rule over nature. However, although political ideas put forward the idea of command and control, the divine and the divine in nature never quite left the frame. Nature then, can be a guide to action.

Nature in a strictly secular sense, it is true, cannot be the source of norms or moral action. Specifically, in this case, nature could never be a guide to make decisions on territorial rights because disenchanted, scientifically measured nature is inert. Nature in the dominant account of secularized modernity is simply the background of culture: "the environment," after all, is what *surrounds* man and his reason. Man remains the source of all moral value in this modern account.

However, nature —and water, specifically— may indeed be a source of normativity if we frame it not exclusively as a resource to be used, but instead as both symbol and proof of the entanglement of humans, animals, plants, and things in place and time. Using this different ecopolitical imagination, we can see in water cycles the precise combination of relations, assemblages, or entanglements that produce coherent socioecological systems. These systems have internal equilibria that can be conceived as a normative goal, and this, in turn, can be a source of moral value in other aspects of political life. Water and the shape of water cycles can help us understand and regulate relationships with ecosystems, and it can be a way to mediate between human communities that want to assert their difference—it can be a tool for asserting plurality and difference without relativism.

For example, water can regulate relationships between humans and other creatures and things around them. Bad relationships with other creatures tend to affect water cycles, and this breakdown leads to damaging droughts and floods. Where forests, rivers, and floodplains have been respected, rivers can overflow their banks without destroying human life. Even without climate change, the behavior of water can say much about the shape of human relationships with the land. Listening to water can allow us to tackle sustainability for both people and ecosystems.

Water can also guide relations among diverse human polities. The shape of water cycles can also help human polities deal with plurality without giving up universal value. Water is an important resource for every single human group, yet we all see water in terms of our cultures. Water is culturally specific, but we all universally agree on its importance. Thus, water can be the proxy for describing the value of which translates into environmental justice. Could water itself guide political decisions on conservation if we imagined the replenishment of the reservoirs as a function of natural cycles that involve plants, animals, and humans in relation to the land where they stand? Could water itself guide political decisions? Could water itself determine who has territorial and resource rights?

Water can make decisions. We could imagine catastrophic drought and flood as water's protest of the disruption of the structures that had previously allowed bodies of water to flow and reproduce. Water cycles point to dynamic equilibria with most living creatures in particular ecosystems. Water could guide political decisions if people seek to restore a dynamic equilibrium that allows human beings to thrive and grow in particular areas together with an idea of order that includes dynamic combinations of plants, animals, and things.

Nature, as a source of value and a guide to political decisions, then, is not strictly speaking "natural" in a secular sense. It is not the sense that pits environment (nature) against culture (the human-made) and culture against grace. "Nature" here does not stand for the scientific background that is distinct from the human world and with its conventional grounding of rights and entitlement. Instead, "nature" here should be interpreted as that force that entangles all things, the very "buzzing" that gives meaning to entangled human action and that that could in turn be a source of ethical grounding for political claims.[25] Granted, just as it did in premodern and early modern thought (which relied on *natural* law), here the hope of relying on nature to make political decisions is an explicit return to political theology. But this political theology of the Earth—as already described—is an alternative to the lingering political theologies of secularized modernity.

This eco-theological alternative, however, is not as readily available as the other two in current political debates. Those eco-theological views that illuminate the complex relations of things, people, and energy over time generally say little about how exactly the relation between individuals and territories ought to be constructed or about who should have jurisdiction over problems and solutions. Perhaps because of this gap between philosophical/theological formulations and political narratives that motivate action, some of the more interesting proposals that could allow us to rethink the relation between individuals and things in political theory have found little uptake among normative political philosophers and legal scholars.[26] Perhaps the place to look for how to motivate political action is not in philosophical discourses that illustrate ideas, but in the experience of communal practices that make water as a source of value.

INSTEAD OF CONCLUSION: AN EXAMPLE FROM THE EXPERIENCE OF MEXICAN PUEBLOS

Just as it was in San Bartolo Ameyalco, the political conflict over water throughout Mexico is often a conflict over experiences. Some experience water as a resource to be used or sold, others see in water the possibility of continuing the connection to the places where they live. For many people "the defense of water and territory" is not always a conflict over water as a resource to be sold, but instead a fight to maintain water as a lifeline for a threatened way of life. Water connects communities to the land, and that connection or grounding in place is what gives value to the land for the people who live there and refuse to migrate to the cities or other countries. Listening to water in solving these conflicts often leads to the idea of *pueblo*: a grounded community.

In most of rural Mexico, grounded communities organize themselves in *pueblos*. "Pueblo" can be translated into English as "the people" or "the town," and it has a special status in traditional Spanish and Latin American legal thinking. A *pueblo* is a corporation with legal personhood that is composed of neighbors or *vecinos*. In the Spanish-American legal tradition, a "vecino" enjoyed a type of territorially bound citizenship based on performance rather than on filial ties or place of origin. *Pueblos*, moreover, are often the inheritors of pre-Hispanic polities—which called themselves *Altepetls* (literally translated from the Nahuatl as "water-mountain": *atl*=water; *tépetl*=mountain).

Pueblos' communal relationships are often tied to divinities (the town's patron saint), but they are also tied to a place and its water. This relationship illustrates how water can help us make decisions. In the contentious politics of water in these grounded communities, we can see how some have navigated the current dualism between those who want to profit privately from natural resources—those who see territories as a means to distribute natural resources for everyone through markets—and those who see water in terms of national (state-based) sovereign self-determination. As an alternative to private-property individualism and state-based development, there are those who see the water as an integral part of the town and its history. In these political conflicts, *pueblos* are often recognized by all parties (including private capital and the state) as political subjects. Even if others don't see them as a legal subject of territorial rights, everybody recognizes that they are a political force that stands as a viable alternative to the individual and the state as a holder of territorial rights. Their political struggles exemplify how water can be the main axis of a form of politics (a form of politics that, I believe, will structure our lives in the future).

The political experiences of *pueblos* illustrate how citizenship can be thought of in terms of the entanglement of people with the intrinsic value of animals, plants, and things. Unlike the main current theories of territorial rights, which see resources as property, the historical experience of *pueblos* does not always justify the grounded community's claim to territory by appeal to individual rights of autonomy, to sovereignty, or to self-determination. Rather, it justifies the claim because of the interconnections of people to nature in specific places and to the place-specific duties that arise from these relations.

Geographical neighbors (including nomads, migrants, and temporary residents) are bound by what I call "place-specific duties": obligations to coordinate with each other regarding the places where they are.[27] These duties spring from physical presence and residence, and they constitute a lattice of dynamic relations among people, biota, and things. This lattice in turn could establish the limits of jurisdictions and determine the scope of political and legal power in each space. On this ground, we could understand wider regions in terms of economic, cultural, and natural networks to which peoples belong and through which they constitute and access resource rights.

According to this way of seeing the experiences of grounded communities, the residents of San Bartolo should have a right to control the spring and the water that has been traditionally part of the town because of the obligations that neighbors owe to each other and because of their duties of stewardship to the aquifer and the surrounding forest and the duties they owe to other people in the watershed. However, this right is limited by the needs of their neighbors downstream and in adjacent towns, and there should be institutions in place to negotiate with them and with the city government. Moreover, to make these rights valid, the town would have an inclusive democratic organization to insure the inclusion of all residents in the vicinity.

The ideas behind this arrangement would be easily recognizable to current residents of San Bartolo, who consider themselves a *pueblo*. However, they would also be aware of the history of colonization, marginalization, and injustices attached to different uses (and abuses) of this model over centuries. Because of its history, the experience of living in a grounded community will differ across *pueblos* and across the world. However, in each grounded community the connections that create a place have water at their core. Thus, water itself can guide polities of all sizes in making decisions aimed at justice.

Who are the grounded communities that feed the forests that in turn make the water that is brought to Cape Town? To Chennai? To Las

Vegas? The nearness of Days Zero throughout the world reminds us of our inevitable entanglement with other creatures and how pursuing valuable relations with them is the only way to ensure that water will be reproduced. Grounded communities are different everywhere, but in each case, their ability to go on will require listening to water as a messenger.

NOTES

1. Josh Holder and Niko Kommenda, "Day Zero: How Cape Town Is Running out of Water," *Guardian*, February 3, 2018. Day Z was averted after severe use restrictions and strong rains in the summer of 2018.
2. Amrit Dhillon, "Chennai in Crisis as Authorities Blamed for Dire Water Shortage," *Guardian*, June 19, 2019.
3. Catherine Keller, *Political Theology of the Earth: Our Planetary Emergency and the Struggle for a New Public* (New York: Columbia University Press, 2018).
4. Rocío González, Mirna Servín, and Alejandro Cruz, "Pugna por El Agua Desata Gran Trifulca en San Bartolo Ameyalco," *La Jornada*, May 22, 2014.
5. A. John Simmons, "On the Territorial Rights of States," *Philosophical Issues* 11, no. 1 (2001): 300–326.
6. David Miller, "Territorial Rights: Concept and Justification," *Political Studies* 60, no. 2 (2012): 252–68; Avery Kolers, *Land, Conflict, and Justice: A Political Theory of Territory* (Cambridge: Cambridge University Press, 2009).
7. Anna Stilz, "Why Do States Have Territorial Rights?," *International Theory* 1, no. 2 (2009): 185–213; Margaret Moore, *A Political Theory of Territory* (Oxford: Oxford University Press, 2015).
8. Simon Caney, "Cosmopolitan Justice, Responsibility, and Global Climate Change," *Leiden Journal of International Law* 18 (2005): 747–75; Cara Nine, *Global Justice and Territory* (Oxford: Oxford University Press, 2012); Chris Armstrong, "Against 'Permanent Sovereignty' over Natural Resources," *Politics, Philosophy and Economics* 14, no. 2 (2015): 129–51.
9. Mathias Risse, *On Global Justice* (Princeton: Princeton University Press, 2013).
10. Anna Stilz, *Territorial Sovereignty: A Philosophical Exploration* (Oxford: Oxford University Press, 2019).
11. Lea Ypi, "Territorial Rights and Exclusion," *Philosophy Compass* 8, no. 3 (2013): 243.
12. Jeremy Waldron, "Property and Ownership," in *The Stanford Encyclopedia of Philosophy*, ed. Edward N. Zalta (2020), https://plato.stanford.edu/entries/property/.
13. Waldron, "Property and Ownership."
14. Simmons, "On the Territorial Rights of States."

15. Rod Nickel and Nia Williams, "Canada's First Nations Seek Bigger Stakes, Profits from Oil Sector," *Reuters*, March 2, 2018.
16. Megan Blomfield, "Historical Use of the Climate Sink," *Res Publica* 22, no. 1 (2016): 67–81.
17. Mathias Risse, "The Human Right to Water and Common Ownership of the Earth," *Journal of Political Philosophy* 22, no. 2 (2013): 178–203.
18. John Locke, *Two Treatises of Government*, ed. Peter Laslett (Cambridge: Cambridge University Press, 1988).
19. Locke, *Two Treatises*, 27.
20. Carl Schmitt, *Political Theology: Four Chapters on the Concept of Sovereignty*, trans. George Schwab (Cambridge, Mass.: MIT Press, 1985).
21. Ernst Kantorowicz, *The King's Two Bodies: A Study in Medieval Political Theology* (Princeton: Princeton University Press, 1997); Carl Schmitt, *The Nomos of the Earth*, trans. G. L. Ulmen (New York: Telos Press, 2006). A third idea of territory as the property of the people in the state is also grounded in a secularized theology that sees the coming into being of self-determining sovereign peoples in terms of progress. This view, associated with Kant, is also prevalent in political theory, but I don't have space to fully develop this view here.
22. Keller, *Political Theology of the Earth*.
23. Catherine Keller, "A Democracy of Fellow Creatures: Feminist Theology and Planetary Entanglement," *Studia Theologica–Nordic Journal of Theology* 69, no. 1 (2015): 3–18.
24. This "having a say" of all creatures and this "experience of entanglement" lead us to process ontology and philosophy. In Whitehead's words, "We find ourselves in a buzzing world, amid a democracy of fellow creatures; whereas under some disguise or other, orthodox philosophy can only introduce us to solitary substances, each enjoying an illusory experience"; Alfred North Whitehead, *Process and Reality: An Essay in Cosmology* (1927–28; repr. New York: Free Press, 1978), 50.
25. Whitehead, *Process and Reality*, 50.
26. Jane Bennett, *Vibrant Matter: A Political Ecology of Things* (Durham, N.C.: Duke University Press, 2010); Diana Coole and Samantha Frost, eds., *New Materialisms: Ontology, Agency, and Politics* (Durham, N.C.: Duke University Press, 2010).
27. I develop the idea of "place-specific duties" into a theory of territorial rights and borders in Paulina Ochoa Espejo, *On Borders: Territories, Legitimacy, and the Rights of Place* (New York: Oxford University Press, 2020).

Democratic Socialism in the USA: History, Politics, Religion, and Theory

GARY DORRIEN

The convention that democratic socialism is hopelessly un-American has become unsettled. A significant portion of the U.S. American electorate no longer tolerates extreme inequality and is committed to holding off the eco-apocalypse bearing down upon us. In Europe, social democracy has created mixed-economy welfare states in which the government pays for everyone's healthcare, higher education is free, elections are publicly financed, solidarity wage policies restrain economic inequality, and ecological health is a high political priority. These achievements have been difficult to imagine, until recently, in the USA. If democratic socialism is about providing universal healthcare, rectifying economic inequality, abolishing structures of racist, sexist, and cultural denigration, and building a peaceable and ecological society, it sounds pretty good to many who have never known anything but neoliberalism and a burgeoning white nationalism.

The USA, for all its predatory history, also has a history of extraordinary movements for social justice. U.S. Americans have long debated two contrasting visions of what kind of country they want to have. Both are ideal types linked to mainstream forms of conservative and progressive politics. The first is the vision of a society that provides unrestricted liberty to acquire wealth, lifts the right to property above the right to self-government, and limits the federal government to military might and safeguarding the power of elites. The logic of this ideal is Right-libertarian or white-nationalist, legitimizing the dominance of the wealthy, the aggressive, the corporations, and aggrieved white people in the name of individual freedom. The second is the vision of a realized democracy in which the people control the government and economy, self-government is superior to property, and no group dominates any other. The logic of this ideal is democratic socialist or Left-progressive, extending the rights of political democracy into the social and economic spheres.

Right-libertarianism is powerful in U.S. American life despite being impossible, setting freedom against democratic equality. Today, white nationalism is the reigning ideology of the Republican Right, fueled by fear and loathing of being replaced. Democratic socialism is supposedly so un-American that it must be called by other names. But it has a rich history in the USA, even by its right name. My book *American Democratic Socialism: History, Politics, Religion, and Theory* (2021) is an expansive argument on this theme. It interprets the intellectual and political history of U.S. American socialism from 1829 to 2020, arguing that the USA has the richest cultural history of democratic socialism in the world and a substantial, interesting, and complex intellectual and political history. The book contends that Christian socialism has been more important in U.S. American democratic socialism than scholarship on this subject conveys and that the craft-basis of the American Federation of Labor (AFL) was a fatal problem for the Socialist Party it never overcame.[1]

The USA did not have a real labor movement. It just had unions, most of them racist, sexist, nativist craft unions that divided workers from each other, fatally truncating the kind of socialism that was possible. In the 1930s, the founding of the Congress of Industrial Organizations (CIO) briefly raised the possibility of a socialist breakthrough, but that was forestalled by Franklin Roosevelt, World War II, and a postwar Congress that outlawed nearly every tool that built the unions. To democratic socialists of my generation, the answer to the labor problem was to treat the Democratic Party as a labor party in disguise, or at least the hope of one. Meanwhile the socialist Left cratered everywhere except one place, the academy, where the Left developed rich new conceptions of social justice emphasizing race, gender, and sexuality as sites of oppression. For thirty years the Left was completely overrun by neoliberal globalization. In 2011, Occupy Wall Street, a spectacular eruption, signaled that many people were fed up with severe inequality and being humiliated. Today, democratic socialist activism is surging as a protest that global capitalism works only for a minority and is driving the planet to eco-apocalypse.

Democratic socialists founded the first industrial unions, pulled the Progressive movement to the Left, played leading roles in founding the National Association for the Advancement of Colored People (NAACP), founded the first Black trade union, proposed every plank of what became the New Deal, and led the civil rights movement of the 1950s and 60s. The best traditions of socialism, I believe, are like the original socialist movement in being predominantly cooperative and decentralized. Nationalization is only one form of socialization and usually not the best one. I believe in expanding the cooperative sector and building bottom-up economic democracy wherever possible, but I also recognize that public ownership at the local, regional, and national levels is sometimes the best option. The convention that democratic socialism is too idealistic to be a viable alternative must be challenged. There had damned well better be an alternative to neoliberalism and destroying the planet.

The USA had vibrant radical democratic traditions before and after Europeans invented socialism. New York disciples of British socialist Robert Owen founded the world's first labor party in 1829, recruiting radical democrats to the view that industries and land should belong to everyone. European socialists poured into the USA after the liberal revolutions of 1848 were put down and socialists had to flee. German American Social Democrats founded the Socialist Labor Party in 1877 along with a smattering of native-born anarchists and Marxists. Christian socialism sprawled across the nation in the 1880s and 1890s, often taking a Populist form. Populists railed against banks and monopoly trusts, calling for silver dollars, founding powerful organizations, seeping into the Democratic Party, and often converting to socialism or Christian socialism.

Very soon after the Socialist Party was founded in 1901, it was a wondrous stew of radical democrats, neo-abolitionists, Marxists, Christians, Populists, feminists, trade unionists, industrial unionists, Single Taxers, anarcho-syndicalists, and Fabians, both American-born and coming from every European nation and Russia. German trade unionists created a powerful Socialist tradition in Milwaukee, where Social Democracy was a culture, not merely a cause. Jewish garment workers from Russia and Russian Poland created similar organizations in New York, espousing a universalistic creed in Yiddish. Rebellious tenant farmers in Oklahoma, red populists in Texas, syndicalist miners in Colorado and California, and populist

Socialists across the Midwest and West built a sprawling network of periodicals, summer camps, and state parties.

The leading Socialist periodical, *Appeal to Reason*, was published in Kansas and topped 900,000 subscribers in its heyday. *The National Ripsaw* morphed out of *Appeal to Reason* and reached a similar audience of farmers, Populists, Christian socialists, and rebels. *The Jewish Daily Forward* was the Bible of New York Jewish socialism, averaging 150,000 subscribers for decades. Scores of Socialist weeklies had upward of 30,000 subscribers, showing that socialism had no trouble speaking American. One of them, the *Texas Rebel*, fairly raged to its 28,000 readers that if you really believe in government of the people, by the people, and for the people, you have to be a democratic socialist; in fact, you are one.

The early Socialist Party was remarkably successful at politics, despite its labor problem, and had little trouble speaking U.S. American, despite its Marxian cast. The first great U.S. American socialist leader, Eugene Debs, was a thoroughly American lover of working-class people who adopted a magical idea of socialist deliverance. The first great hope of radical industrial unionism, the Knights of Labor, was founded by Christian socialists. It got pulled into more strikes than it could handle and learned bitterly that state governments stood ready to smash them. In the USA, unionism mostly meant craft unionism, which organizes the workers of a specific skilled job. The overwhelmingly craft basis of the AFL fatally truncated the labor movement and the kind of socialism that was possible, thwarting socialists from scaling up and from creating a labor party.

The USA established a simple-majority system in which a party receives no representation if it does not win more votes than all other parties within a constituency. Simple-majority representation in single-member districts puts immense lesser-evil pressure on voters not to waste their vote. The plurality model of representation turned the nation into a two-party fiefdom that thwarted third-party challenges. There are almost no exceptions in the world to the rule that a simple majority single-ballot system creates two-party fiefdoms. Many have argued that U.S. American socialists would have floundered anyway under a system of proportional representation because socialism was no match for America's open borders, prosperity, and upward mobility. Many U.S. American workers feared that socialism would prevent them from getting ahead. But the USA had more than enough suffering and exploitation to create a surging socialist movement.

The number-one problem for U.S. American socialists was that divide-and-conquer worked in the USA.

Workers were turned against each other, pitting native-born workers in the craft unions against unskilled immigrant workers. The AFL bought into capitalism and excluded the mostly immigrant industrial workers. The Socialist wing of the AFL, consisting mostly of five industrial unions in the mining, brewery, and garment industries, plateaued at 38 percent of the AFL. That yielded a labor movement unlike any in Europe, a crushing difference for U.S. American socialists. No factor outranked this one. U.S. American unions were founded separately from left-wing political parties. They protected their independence from all political parties, becoming part of the system of political control represented by the two-party system, and defeated the Socialist union leaders who stumped for a labor party. Karl Marx perceived that this exceptional characteristic would be difficult to overcome. Then Eugene Debs condemned the Socialist comrades who tried to win the AFL to socialism, spurning them as reformist sellouts. Debs played a role in sealing the greatest failure of the socialist movement by charging, justly and vehemently, that the business unionism of the AFL made it beholden to the interests, worldview, and agenda of the ruling class.

The Socialist Party peaked at 118,000 members in 1912, which sounded impressive only in the USA; that year the British Labour Party boasted 1.9 million members. There were never enough unionists in the Socialist Party or industrial unionists in the AFL to sustain socialism. Craft unionism so dominated the AFL that craft racism and sexism were impregnable and political independence was orthodoxy.

Debs would not have become a socialist had he been able to tolerate remaining in a craft union. His fling with industrial union leadership was stormy and brief, after which he converted to magical socialism, the cure for all social problems not to be sullied by reform movements or mediocre trade unions.

Debs was the apostle of a true way that found strength in its evangelical purity. His socialism was a Protestant redemption strategy soaked in the idioms and assumptions of American revivalism. Being a romantic U.S. American individualist reinforced his magical socialism and his evangelical concept of his mission, making him an incomparable platform performer. He loved the workers, and they loved him back, but he made it hard for them to join his party, and he spurned the strategy that worked in England—forming a coalition party of the democratic Left.

This wondrous Debsian socialism was destroyed in 1917 and 1919. The Socialist Party bravely opposed World War I and paid a horrific price for it, viciously persecuted by the government. Then the meteor of world Communism crashed into the Socialist Party and blew it apart. The Debsian heyday ended in shattered despair, yielding the dismal run-up to "Norman Thomas Socialism," as it was called. Norman Thomas, a Presbyterian minister who graduated from Union Theological Seminary, joined the Socialist Party in 1917 because it opposed U.S. intervention in World War I and the Presbyterian Church did not. He quickly rose to the top of the party because most of the party's native-born intellectuals fled to Woodrow Wilson, and Thomas offered a noble contrast to patriotic gore. Norman Thomas Socialism was a three-sided struggle to renew the democratic socialist idea, hold off the Communist Party, and get a farmer-labor-socialist-progressive party off the ground.

The industrial unions always played the leading role in pushing to create a labor party. In 1920 they struck out on their own, founding the Farmer-Labor Party. The ascending British Labour Party inspired them, and the Communist breakup of the Socialist Party repelled them. The first national Farmer-Labor party made a dismal beginning, but four years later the forces that needed to come together briefly did so, for one election, running Wisconsin Progressive U.S. senator Robert La Follette for president. It helped that the AFL came aboard to punish the Democrats and Republicans, but the AFL had not changed; backing La Follette was a one-off affair. The dream of a labor party stayed out of reach, condemning the Socialists to years of irrelevance kept afloat by garment union money.

The farmer-labor-socialist-progressive coalition was never hard to imagine. It haunted the Left because every election produced political victors who did not represent vast sectors of the population. W. E. B. Du Bois, Reinhold Niebuhr, John Dewey, and other intellectuals in Thomas's orbit shared his dream that the disparate Left would pull together. In 1935, Thomas dragged the Socialist Party into solidarity work with the fledgling Southern Tenant Farmers' Union in Tennessee. The union grew rapidly and adopted a Black church hymn, "We Shall Not Be Moved." Thomas risked his life by speaking to terrorized sharecroppers in Arkansas. He pleaded for a meeting with Agriculture Secretary Henry Wallace, who refused to see him. Thomas despised Wallace for the rest of his life, which was fateful in the mid-1940s, when Wallace became a leading anti-anti-Communist and Thomas spurned him for that reason, too.

For a while the Great Depression rewrote the script on what might be possible. Union activism rebounded dramatically, Congress passed the Wagner Act of 1935, and Communists and Socialists organized the CIO. The Wagner Act threw the weight of government behind union organizers, forcing employers to allow their plants to be unionized. Franklin Roosevelt endorsed it shortly before it passed, co-opting a tide of left-wing and right-wing populist forces—Norman Thomas Socialists, Farmer-Labor organizers, Huey Long fascists, Republican progressives, and Communists. He did it with wily brilliance, putting Leftist leaders on his payroll, favoring select third-party candidates over Democrats, telling them he was on their side—determined to transform the Democrats into a progressive party. The New Deal enacted 90 percent of the Socialist platform; to a considerable degree, the New Deal was a form of socialist deliverance. The Socialist play could have been to pull FDR to the left by working with him and demanding more from him. But the Socialists opposed him, clinging to socialism-is-the-answer, which made them look irrelevant.

Thomas was eloquent, personable, astute, courageous, and not cut out to be a party leader. He symbolized the shift of the Socialist Party from being primarily working class to being primarily a vehicle of middle-class idealism. The New York garment unions were the financial rock of the party until 1937, when Thomas and the left wing drove them out. Afterward there was no financial rock. For forty years, Thomas and Black socialist union leader A. Philip Randolph stood together at the center of democratic socialism. The party's united front activism mostly backfired, and the party dwindled, surpassed even by a Communist Party that was shrewd enough to support Roosevelt. Thomas and the Socialists allowed into the party a band of Trotskyites who sabotaged the party and stole its youth section. Another exodus ensued when Thomas and the Socialists held out too long against World War II. Afterward, Thomas adamantly opposed Soviet domination of East Europe and pro-Soviet American Leftism, supporting the purge of Communists from CIO leadership positions.

The last hope of a Labor Party was lost in the whiplash reactions of 1946–48. The CIO struck hard for postwar wage gains, and Congress passed the Taft-Hartley Act over Harry Truman's veto. Taft-Hartley abolished or curtailed almost every tool that built the unions, outlawing jurisdictional strikes, wildcat strikes, solidarity strikes, secondary boycotts, secondary and mass picketing, closed shops, and union contributions to federal political campaigns. It gave state legislatures a green light to enact Orwellian right-to-work laws having nothing to do with the right to work. The unions had grown from three million AFL members in 1935 to fourteen million AFL and CIO members in 1945. Taft-Hartley was about making them weak and insecure again. The last hope of a Labor Party died with Truman's feisty comeback victory of 1948. Now the defanged labor movement belonged wholly to the Democratic Party, an outcome facilitated by the merger of the AFL and CIO in 1955.

Thomas's pilgrimage from social gospel socialism to the leadership of the Socialist Party exemplified a Christian socialist tradition that has never gotten its due in the literature on U.S. American socialism. The lack of interest by scholars in Christian socialism has yielded accounts that do not explain how African Americans and feminists came into the movement through religious socialism. Two classic histories of democratic socialism published in 1952 dominated this field for a generation, summarizing opposite traditions of assessment yielding a similar verdict.

Political scientist Ira Kipnis, in *The American Socialist Movement*, argued that the Socialist Party was doomed from the beginning by its accommodating of social democratic reformism. Kipnis said the right wing of the party, led by Milwaukee journalist-politician Victor Berger, was consumed with winning elections, and the mainstream of the party, led by New York journalist-politician Morris Hillquit, was only slightly less opportunistic. The party debated immediate demands and true Marxism at its founding but adopted the wrong answer. It got a second chance at correcting its course in 1905, when the Industrial Workers of the World (IWW) was founded, but Debs did not stick with the IWW, and most socialists loathed its anarcho-syndicalism and violence. The party lost its last chance of becoming important when it censured its left flank of IWW members in 1912 and expelled IWW leader Bill Haywood from the executive committee the following year. Kipnis contributed mightily to the legend that the IWW Wobblies were the real thing and the Hillquit-Berger socialists were sellouts. The real thing was anarchist in hating government and syndicalist in contending that worker syndicates should run the country. True leftism versus opportunism explained the failure of the Socialist Party, culminating in the Haywood drama.

Sociologist Daniel Bell, in *Marxian Socialism in the United States*, agreed from an opposite standpoint that the party was hopelessly futile from the beginning, contending that every Socialist leader espoused a utopian vision of social transformation that made the party alien to American society and marginal in it. According to Bell, AFL leader Samuel Gompers was the wise hero who figured out how to make social democratic gains in capitalist America, whereas Debs, Hillquit, and Berger clung to an un-American fantasy. Subsequently, the Socialist Party crawled onward at the national level only because it had a compelling figurehead, Thomas. Just as Kipnis looked down on his subject from the superior vantage point of pro-Communist radicalism, Bell looked down as a Cold War liberal, having recently outgrown his youthful attachment to Norman Thomas Socialism. The socialists, Bell said, were ideologues in a pluralistic and technocratic society that eventually put an end to ideology itself.

These rival books cast a long shadow over scholarship on the American Left. The fact that they drove to the same conclusion about the futility of democratic socialism solidified this verdict as a convention. Kipnis and Bell summarily dismissed the Christian socialists who spoke to broader middle-class audiences than the Socialist parties, demanded to be included in socialist politics, and built significant organizations. Kipnis wrote them off in three quick strokes, noting that George Herron was briefly famous, something called the Christian Socialist Fellowship existed, and all Christian socialists, being religious, were of course opportunists. Christian socialism itself he dispatched with a single sentence: "Since the Christian Socialists based their analysis on the brotherhood of man rather than on the class struggle, they aligned themselves with the opportunist rather than the revolutionary wing of the party." The party's many Christian socialist leaders and authors, whoever they were, could not have mattered, since they were religious.[2]

Bell similarly pushed aside the Christian socialists without employing "opportunist" as a broad-brush epithet. He devoted a footnote to the Christian Commonwealth colony at Commonwealth, Georgia, noted that Edward Bellamy's Fabian utopian fable *Looking Backward* (1888) won most of its fame through Christian socialist clergy, and observed that a cleric named George Herron was "one of the leading figures of the party." That was it. Even a bit of following up on Herron would have vastly enriched Bell's picture of U.S. American socialism, but he wasn't interested. It could not be that these people mattered. The struggles for racial justice and feminism had no role in Bell's story, so the Christians in them didn't matter, either. Bell's very insistence that socialism is always religious—that is, eschatological—exempted him, he thought, from paying attention to any actual religious socialists, whether or not they were indebted to Marx.[3]

Herron was a lecture circuit spellbinder and Congregational cleric who befriended Debs, showed Debs how to translate ethical idealism and populism into sermon-style socialist evangelism, and electrified the social gospel movement by calling America to repent of its capitalist, racist, sexist, and imperialist sins. W. D. P. Bliss was a tireless organizer and Episcopal cleric who tried to unite the reform movements and failed to persuade the Socialist Party that uniting the reform movements was its mission. Woodbey was a brilliant Black Baptist cleric who spoke for the Socialist Party and the IWW, was beaten and jailed for doing so, and tried to improve how the party and the Wobblies talked about racial justice. W. E. B. Du Bois had one foot in the Black church, joined socialists Mary White Ovington and William English Walling in willing the NAACP into existence, and provided intellectual leadership for Reverdy Ransom, Robert Bagnall, George Slater, George Frazier Miller, and other Black social gospel socialists. Walter Rauschenbusch, the leading social gospel socialist of his time, never quite joined the Socialist Party because he recoiled at its atheist officials. Kate O'Hare was a brilliant prairie socialist writer and speaker who reflected the racism of her milieu and attracted a following exceeded only by Debs. Vida Scudder was a prolific organizer, writer, Episcopal laywoman, feminist, and lesbian who worked with Bliss and tried to drag Rauschenbusch into the Socialist Party.

These apostles of Christian socialism absorbed more Marxist theory than they usually found it prudent to cite. Bliss and Herron were like Debs in coming to socialism through the Populist movement and its outraged moral sensibility. Bliss, Herron, Scudder, and Rauschenbusch struggled with the paradoxes of their ethical Christian idealism for socialist activism, but like Debs, they believed that the class struggle and the limits of middle-class idealism compelled them to be socialists. They said so eloquently a generation before Reinhold Niebuhr became famous for saying it. Marxian social democracy and Populism were the two main highways into

American socialism. Christian socialism was the third, and much of the Populist movement *was* Christian socialist. The Niebuhr generation of Christian socialists included Mordecai Johnson, Walter Muelder, Kirby Page, Sherwood Eddy, J. Pius Barbour, and Benjamin E. Mays. They took for granted that the best forms of Christian theology and ethics are Christian socialist, passing this conviction to Martin Luther King Jr.

In 1958 Thomas reluctantly admitted a group of former Trotskyites into the Socialist Party, fearing they would take it over, which they promptly did. These were the Shachtmanites, disciples of Max Shachtman, a former associate of Bolshevik hero Leon Trotsky. The Shachtmanites were brainy, cunning, scholastic, aggressively parasitic, fiercely ideological, consumed with the right kind of anti-Communism, which they called anti-Stalinism, and at every historical turn, strange. They were still Leninists when they broke from Trotsky in 1939 and were more Leninist than they claimed when they morphed in the mid-1950s toward democratic socialism. They found Thomas boring and Shachtman exhilarating. Michael Harrington was their youthful star. Brilliant, energetic, and charming, he befriended Black socialist-pacifist organizer Bayard Rustin and brought Shachtmanites into the civil rights movement.

For most of its history, the Socialist Party took a decent position on racial justice and did little about it, falling short of the Communist Party. Randolph, Rustin, Harrington, James Farmer, Ella Baker, and former Communists Stanley Levison and Jack O'Dell changed this picture, helping King unite the established civil rights movement based in New York City with the new, youthful, church-based movement of the South. Rustin and Harrington organized key civil rights demonstrations of the late 1950s and early 1960s, while Rustin joined the Shachtmanites. In 1960, Harrington and Rustin helped the Shachtmanites overtake the Socialist Party. Both were dedicated to keeping secret that King's social gospel was socialist. Harrington was anointed the successor to Debs and Thomas, a title he didn't deserve until he broke from the Shachtmanites in 1972 and broke up the Socialist Party.

The Shachtmanites had a vision of a realigned Democratic Party that enacted the agenda of the AFL-CIO, supported the civil rights movement, and drove out the party's Dixiecrat flank. They were done with the warhorse doctrine that socialists should never ally with bourgeois parties. The Democratic Party, they claimed, was becoming a labor party in disguise.

Shortly after the Shachtmanites swung the Socialist Party behind this strategy, a group of ambitious college students based in Ann Arbor, Michigan, proclaimed that a "New Left" was needed. The leaders of Students for a Democratic Society (SDS) lumped together all the competing groups and ideologies of what they derisively called the "Old Left." Thomas got a pass, as did Harrington at first, but the SDS said it took no interest in Old Left fights over Marxian ideology, Communism, unions, and the working class. Anti-Stalinist social democrats were surely better than pro-Soviet Communists, but only by degree. To the SDS, the Old Leftists sounded too much alike, not fathoming what it was like to be a college student in 1962.

The New Left was born in a fractious relationship with the Socialist Party while depending on funding from trade unions in the party. The so-called Old Left, being cast as old and bygone, denied that privileged college students who never learned their Marxism had anything to teach them. The socialist drama of the early 1960s pitted hardened survivors of the 1930s against gently raised youth of the 1950s. It built to a spectacular crash as the SDS self-imploded, leaving the Old Left socialists to say I-told-you-so. The Black New Left struggled with the role models it inherited from the 1950s while the white New Left was too alienated to find any; social critic C. Wright Mills came the closest to being a half-exception. The New Left wrongly spurned the hard-won wisdom of the Old Left about Communist tyranny, but it gave birth to liberation movements that enriched how socialists conceived social justice and battled for it. Harrington blew his chance to be a bridge figure between the Old Left and New Left—until the 1970s.

The 1970s was a lost decade in U.S. American politics that absorbed the turbulent legacy of the 1960s and the daunting transformation of the world economy. The economic boom of the post–World War II era ran out, yielding a structural economic shift and its miserable combination of stagnation and inflation. Stagflation defied Keynesian correction, confounding the social democratic Left. The bitter ideological divides in the Socialist Party blew it apart in 1973, ending the party of Debs and Thomas. The Shachtmanites bridled at the anti–Vietnam War movement, Black Power, and radical feminism, founding Social Democrats USA. Harrington led a faction of progressive social democrats into a new organization called the Democratic Socialist Organizing Committee (DSOC), building a vehicle for Old Left social democrats, select veterans of the New Left, and youthful newcomers from George McGovern's Democratic presidential campaign. Meanwhile, Harrington argued that the rightward trajectory of the Shachtmanites represented something too important not to name. He called it neoconservatism, a tag that stuck. The Shachtmanites and Cold War liberals he named went on to become the most consequential intellectual-political movement of their time, winning high positions in three Republican administrations and mocking Harrington for befriending feminists and anti-anti-Communists.

The idea of the DSOC was to create a multi-tendency organization uniting the generations of the progressive democratic Left. The DSOC was more Old Left than New Left, wearing its anti-Communism proudly. Yet the DSOC achieved the Communist Party dream of the Popular Front periods of 1935–39 and 1941–45, creating a united front organization, this time without Stalinism. The DSOC won the battle against the neoconservatives for influence in the Democratic Party only to get blown away by the next great turn in U.S. American politics. Harrington and the DSOC sought to ride into power in 1980 when their ally Edward Kennedy challenged Jimmy Carter for the Democratic nomination. Instead, Kennedy failed to unseat Carter, and the neocons rode into power under Ronald Reagan. The DSOC was too deflated by defeat and disdainful of Carter to rally for him against Reagan. Many blamed the hapless and unlucky Carter for the alarming triumph of the Reagan Right, but Harrington stressed that Reagan became powerful by offering clear, bad, popular answers to complex problems. The Left needed new answers calibrated to the new realities of global capitalism.

In 1982, the DSOC merged with a New Left organization, the New American Movement, to form the Democratic Socialists of America (DSA). There was no mistaking the symbolism of the DSA—it was founded to heal the leftover rift between the Old Left and New Left. The DSA debated the fiscal crisis of the state and two academic cottage industries called "market socialism" and "analytical Marxism," but these were sideshows compared to the rise of a cultural Left that emphasized race, gender, and sexual identity as sites of oppression. Not coincidentally, a long-departed Italian Communist leader, Antonio Gramsci, won a tremendous vogue for contending that the Left wrongly cedes the entire cultural realm to the Right.

Gramsci died in a Fascist prison cell in 1937. He argued that capitalism exercises "hegemony" over the lives of people where they live in schools, civic organizations, religious communities, newspapers, media, and political parties. Hegemony is the cultural process by which a ruling class makes its domination appear natural. Gramsci contended that if the Left had any serious intention of winning power, it had to contest the Right on the cultural level. This argument swept much of the socialist Left in the 1980s, providing socialists with a sort of Marxian basis for appropriating the cultural leftism of identity politics, difference feminism, and other forms of cultural liberation.

The academy had never played an important role in the socialist Left until socialists from my generation embarked on academic careers. I was a holdout from the surge into the academy, having worked as a solidarity activist in democratic socialist and anti-imperialist organizations before I became an academic at the age of thirty-five in 1987. By the time that I entered the academy, I was well behind the career trajectory of my academic friends and still surprised to be there. The socialist Left cratered everywhere except the academy, where some on the cultural Left lifted recognition claims above economic justice, some old-style social democrats inveighed against the rise of cultural Leftism, and Cornel West, bell hooks, Iris Marion Young, and Nancy Fraser differently made seminal arguments for fusing redistributionist and recognition politics.

West combined Black liberationist, Council Marxist, pragmatic, and Christian socialist perspectives, developing a formative socialist theory of racism and surpassing all others of his generation as a Gramscian public intellectual. hooks was a pioneer of Black socialist feminism and the Black feminist tradition of intersectional analysis, conceiving race, gender, sexuality, and class as interlocking variables not reducible to hierarchical ordering. Young contended that the Left needed a concept of justice that emerged from listening to liberationist movements, not from applying abstract principles of justice to society. She developed a fivefold concept of oppression as exploitation, marginalization, powerlessness, cultural imperialism, and violence, stressing that distribution is always at issue in these forms of harm and that none is reducible to distribution. Oppression happens to entities that no theory of justice has ever conceptualized—social groups, which are socially prior to individuals without existing apart from individuals. Equal treatment, the gold standard of fairness theories of justice, suppresses differences in ways that reinforce oppression.[4]

These arguments were hotly contested on the Left while the political Right waged its loud attack on the welfare state and Bill Clinton demoralized the Left by carrying out Democratic versions of Republican policies. Communitarian theory flourished during this period. Communitarians ranging from democratic socialists (Benjamin Barber, Robert Bellah, Rosemary Radford Ruether, Philip Selznick, William Sullivan, Michael Walzer), to moderate progressives (Amitai Etzioni, William Galston, Jane J. Mansbridge, Michael Sandel, Charles Taylor), to conservatives (William Bennett, Alasdair MacIntyre, Robert Nisbet, Christina Hoff Sommers) criticized the egocentrism of U.S. American culture and the liberal devotion to individual rights. They revived the entire field of political theory, retrieving Aristotle's concept of justice as a community bound by a shared understanding of the good and Hegel's emphasis on recognition. But the communitarians made little impact on the Left, where their rhetoric of family-community-nation smacked of conservative piety.

Nancy Fraser made a landmark case for a fusion of socialist redistribution and cultural recognition while sharply rejecting Young's optimism about the complementarity of these orientations. Fraser said it was wrong for the Marxist/social democratic Left and the difference-feminist/multicultural Left to fight over the hierarchy of oppression, a mistaken debate with harmful consequences. The major axes of injustice, she argued, are two-dimensional. Every form of injustice is rooted simultaneously in the political economy and the status order. No struggle for justice can succeed lacking a politics of redistribution *and* a politics of recognition. The hard part comes next, because these two orientations are not complementary. Redistribution strategies silence the most pressing causes of harm for denigrated groups, while recognition strategies mitigate unjust outcomes without changing the economic structures that generate unjust outcomes. Moreover, the distributive justice of the welfare state and its multiculturalist approach to cultural harm are both inadequate. Fraser urged the Left to combine socialist redistribution with difference feminism and cultural deconstruction.[5]

Cultural accounts of injustice are symbolic, rooting injustice in social patterns of representation and interpretation. Here the defining injustices are disrespect, being rendered invisible, and being judged by cultural norms that are alien to one's culture.

Late capitalism decentered the importance of class, after which social movements mobilized around crosscutting axes of difference. Fraser stressed that the conflicts between socialist redistribution and cultural liberation stripped the Left of its former coherence. Recognition politics promotes group differentiation by advocating for specific groups, while redistribution politics seeks to abolish group differentiation.

Fraser devised a social spectrum bordered at one end by the redistribution model and at the other end by the recognition model, construing gender and race as hybrid modes in the middle, combining features of an exploited class and an oppressed sexuality. Both forms of injustice are primary and co-original. She ended up with a four-celled matrix placing redistribution and recognition at opposite ends of a vertical axis and affirmation and transformation remedies at opposite ends of a horizontal axis. Affirmation remedies operate within the system; transformation remedies abolish it. Two combinations came out better than the others. The welfare state meshes with multiculturalism, since both are affirmation strategies. Democratic socialism and cultural deconstruction also go together, since both are transformation strategies. Fraser argued that combining socialism with cultural deconstruction is the only way to do justice to all struggles against injustice. Affirmation strategies assume a zero-sum game and do not promote coalition building. The conflict between redistribution and recognition is especially acute across collectivities such as gay and working class or Black and female. Affirmation strategies work additively and conflict with each other. Transformation strategies try to promote synergy, not being zero-sum.

The Fraser debate carried on for over a decade. The second round was grim, chastened, and sometimes despairing. It operated within Fraser's dual framework and took for granted that the chasm on the Left was real and deep. Fraser sharpened her critique of chauvinist elements within recognition movements and blasted Young for idealizing the cultural Left. A third round of debate, commencing in 2002, challenged Fraser's economy-culture model, contending that it left no room for the vital political sphere of law, citizenship, and institutions. The political dimension supersedes redistribution and recognition because it is normatively and conceptually prior to other forms of social participation. Fraser refashioned her theory in response, now treating the problems of political representation as justice concerns. Redistribution and recognition are political in contesting for power and objectives, but politics determines how struggles for justice are structured.⁶

Fraser made a historic contribution to democratic socialist theory. She rightly contended that whatever the organizing frame of global politics is going to be beyond the Westphalian nation state—which nobody knows—it must begin with the socialist principle that all who are affected by a given structure or institution should hold moral standing as subjects of justice in relation to it. Today, in my view, there are two fronts of the struggle to achieve the principle of all-affected moral standing. One is the prosaic political struggle to secure the right to vote and attain decent government policies. The other is the global fight led by people of color, environmentalists, indigenous peoples, feminists, and solidarity activists to claim their standing as subjects of justice. There is a right to make a claim of injustice against any power that causes harm. At least, there should be. Meanwhile, I do not accept Fraser's contention that affirmation strategies must be left behind. The Right is out to destroy the welfare state, affirmative action, and multicultural education. I am not for helping it in any way, just as I am not the kind of democratic socialist who looks down on social democracy. Germany, Sweden, Denmark, and Norway have high wages, strong unions, free education, free healthcare, monthly stipends to undergraduates, the highest rates of happiness and good health in the world, up to 480 days of paid leave when a child is born or adopted, serious efforts to convert to a green economy, and vibrant economies that are one-fourth publicly owned. Germany has 50 percent worker co-determination on every company supervisory board. I am for as much of that as we can get in the USA.

Bernie Sanders inveighed against corporate greed and inequality for decades before mass movements for social justice were possible again. In 1990 he won Vermont's lone seat in the U.S. House of Representatives as an independent democratic socialist. In 2006 he moved up to the U.S. Senate, already forging a career lacking any parallel in Left politics. In December 2010, Sanders held forth on the Senate floor for eight and a half hours. He had no prepared text; he had only scraps of various speeches and a determination to see how long he could last. All were wrapped around a basic storyline. In the 1970s, he observed, the top 1 percent of earners took home 8 percent of all income. In the 1980s they earned 12 percent. By the end of the 1990s they were getting 18 percent. By 2007 they were up to 24 percent. Sanders pleaded, "How much more do they want? When is enough, enough? Do they want it all?" Greed is a sickness, he said, much like addiction. The 1 percent is addicted to greed: "I think this is an issue we have to stay on and stay on and stay on."[7]

Sanders has the virtue of relentlessly staying on. In 2015 he challenged Hillary Clinton for the Democratic nomination for president. He ran the greatest political campaign ever waged by a U.S. American democratic socialist, winning twenty-two primaries and caucuses. Sanders describes democratic socialism as the belief that a living wage, universal healthcare, a complete education, affordable housing, a clean environment, and a secure retirement are economic rights. He got through the entire campaign without being asked about worker ownership or public ownership, which was fine with him. He was content to fight for economic rights that social democrats achieved in Europe a half century ago. But Sanders is more radical than any European social democratic leader of the past generation because he renewed the language of the class struggle, a language not spoken in Europe or the USA since the 1950s.

His first run for president set off a membership gusher in the DSA that is still climbing six years later. The DSA had 7,000 members when Sanders first ran for president. Then Donald Trump won the presidency, DSA members Alexandria Ocasio-Cortez and Rashida Tlaib were elected to Congress in 2018, and Sanders ran again for president in 2020. Now the DSA was up to 60,000 members. Sanders terrified the party establishment by tying for first place in Iowa, winning the New Hampshire primary, and crushing the field in the Nevada caucus. The Democratic establishment and corporate media shrieked with sky-is-falling alarm, pleading that regular Democrats and Wall Street Democrats had to consolidate before Sanders ran away with the nomination. South Carolina was next, fortunately for Joe Biden. He had never won a primary in three presidential nomination campaigns until he vanquished Sanders in South Carolina. The waters parted for Biden as four moderate candidates and one progressive candidate dropped out, clearing his path to the nomination. Fear of Sanders and fear of a Trump reelection drove the field to consolidate with breathtaking speed.

Today, Sanders rues that his thriving campaign was throttled practically overnight. But the causes he cares about are more prominent than ever in U.S. American politics, and a burgeoning DSA—with 100,000 members—is confronted with questions it never had previously about how to leverage its capacities. The rush of new members into the DSA includes many post-Trotskyites and semi-anarchists who clash with each other over ideology while agreeing that the DSA should disavow its social democratic legacy and reinvent itself as a working-class organization. Others support a renewed focus on working-class organizing without agreeing to break with the DSA's social democratic legacy or its usual practice of working in the Democratic Party.

The DSA has long said that social movement work and public socialist education are its top priorities, whereas electoral politics is important for some and not for others; it can mean different things, and some are outright against it, so it is number three. That has not changed, but the kind of electoral and labor activism that the DSA will support in the future is very much a contested matter. The DSA has caucuses that are class-first and caucuses that are fusionist in the varied fashion of West, hooks (who died in 2021), Young (who died in 2006), and Fraser. I am in the latter group, believing that the USA's original sin of colonial devastation, slavery, and white supremacy must be addressed as the highest priority no matter how much one may believe in a Marxian perspective.

The DSA, across the boundaries of its current debates, is focused on local politics and local labor organizing across the nation, creating chapters that build their own field and canvassing operations, maintain their own data, formulate their own messaging, develop their own research capacity, and even run their own campaigns, acquiring the full range of movement skills and capacities. This contemporary movement is not like the previous generations of DSA activists, for whom socialist activism was secondary to other activist priorities even as they touted its interconnectedness to everything else. The millennials and Generation Z activists who have poured into the DSA expect democratic socialism to be their top priority. They expect to find enough in it to sustain them. They include many of the best organizers in the nation. And there are far too many of them to melt away.

NOTES

1. Gary Dorrien, *American Democratic Socialism: History, Politics, Religion, and Theory* (New Haven: Yale University Press, 2021). This chapter is a capsule summary of arguments and themes in this book.
2. Ira Kipnis, *The American Socialist Movement, 1897–1912* (New York: Columbia University Press, 1952), 269.

3. Daniel Bell, *Marxian Socialism in the United States* (1952; repr. Princeton: Princeton University Press, 1967), 61.
4. Cornel West, *Prophesy Deliverance! An Afro-American Revolutionary Christianity* (Philadelphia: Westminster Press, 1982); Cornel West, *Prophetic Fragments: Illuminations of the Crisis in American Religion and Culture* (Grand Rapids, Mich.: Eerdmans, 1988); bell hooks, *Ain't I a Woman: Black Women and Feminism* (Boston: South End Press, 1981); bell hooks, *Feminist Theory: From Margin to Center* (Boston: South End Press, 1984); Iris Marion Young, *Justice and the Politics of Difference* (Princeton: Princeton University Press, 1990).
5. Nancy Fraser, "From Redistribution to Recognition? Dilemmas of Justice in a 'Postsocialist' Age," *New Left Review* 212 (July/August 1995): 68–93; reprinted in Fraser, *Justice Interruptus: Critical Reflections on the 'Postsocialist' Condition* (New York: Routledge, 1997), 11–39.
6. Kevin Olson, ed., *Adding Insult to Injury: Nancy Fraser Debates Her Critics* (London: Verso, 2008); Nancy Fraser, *Scales of Justice: Reimagining Political Space in a Globalizing World* (New York: Columbia University Press, 2009).
7. Bernie Sanders, *The Speech: On Corporate Greed and the Decline of Our Middle Class* (New York: Nation Books, 2015).

∽ Regifting the Divine Economy: Transitioning Petroleum-Based Energy Regimes

MARION GRAU

RECONSIDERING GIFT ECONOMIES

A great longing is upon us, to live again in a world made of gifts. I can scent it coming, like the fragrance of ripening strawberries rising on the breeze.[1]

Petroleum has functioned as a kind of sacrament for U.S. petro-energetic regimes and fueled neoliberal economic and consumptive structures. The present "global kleptocracy" would not exist in this form without its oily tentacles,[2] and any chance at a just energy transition depends on humans able to narrate change in ways that incite joy. It is therefore to the intersection of energies, exchanges, and fluids that I would like to return to in this chapter, and in particular to the energies and exchange relations of the gift.

In the struggle against the disintegration of U.S. democratic structures, many have exclaimed "this is not who we are," while others have reminded their fellow citizens that "this is exactly who we are"—a nation constituted in and through indigenous displacement and genocide, racial resentment, and greed, while waving the banner of Christian nationalism. After all, the U.S. constitution enshrined certain colonial economic creeds that imposed this settler nation on the lands and peoples that lived there already. These principles have not been effectively purged, nor hardly questioned effectively, but are being reestablished before our eyes in toxic white male oligarchic garb, seeking to establish permanent white supremacist rule by voter suppression and other schemes to disenfranchise people. Trump was lifted to the presidency with the help of a corporate oligarchic culture hell-bent on prolonging the reign of petroleum and the sway it has over U.S. and global energy and political regimes. If democratic structures and climate demonstrators stood against that, so be it. The pattern now ripping apart the last patches of democracy sewed on the garment of colonial indigenous genocide and the Atlantic slave trade is a staple of U.S. economic culture:

> The third millennium began with seemingly countless revelations of counterfeit accounting, insider trading, artificial stock inflation, and other financial scandals that were hushed, if not encouraged by the economies of exchange in a market dominated by stockholder capitalism. As they continue to be exposed, the hucksters and shams of this age oddly resemble those of the 1830s, the Jacksonian robber barons and self-made financiers of the Gilded Age. . . . [Lewis] Hyde suspects that "con men" such as these "embody things that are actually true about America but cannot be openly declared (as, for example, the degree to which capitalism lets us steal from our neighbors, or the degree to which institutions like the stock market require the same kind of confidence that criminal con men need)."[3]

The Trump presidency represented a kind of apotheosis of a known con man and liar. Much could be said about the con man regimes and adulation of the hyper-rich and "billionaires" plaguing our societies. I received my PhD from Drew, an institution with the motto "freely you have received, freely give" (Mt 10:8). It still mocks me with its impossible biblical request. Daniel Drew, the robber baron after whom the university was named, was known for his con man antics and later bankruptcy. The story circulating on campus was that he promised money to the university that never materialized, though it kept his name. Some can therefore see the name Drew as a tragic reminder of faith put in toxic men that promise us fantastic fortunes and futures while speculating on ours and others' confidence.[4]

We are experiencing the "push back of imperious property against democracy"[5] that has allowed anti-democratic oligarchic movers and shakers in the U.S. to paralyze institutional structures that aim for more equitable distribution of power and resources. While the more obvious result is a plutocratic presidency, the economic drive to oligarchy, to limitless greed hiding behind free market phrasings, seems endemic to the colonial origins of the U.S. and its global influence in capitalist economic structures. This form of plutocracy dominates workers and production sites to rig monetary flows toward the pockets of the few beneficiaries and scores of complicit actors. Ultimately, humans and animals on the planet are expendable to that type of economic greed; we are all "externals" that can be excluded from the balance sheet. Each day the violence against ecosystems inflicts unimaginable and unseen forms of suffering against animals and humans, especially non-white and poor people.

Contemporary neoliberal capitalism has been built on the energy regime of petrocapitalism.[6] In many parts of the world, oil extractivism depends on the colonial theft of land from indigenous peoples and caretakers. Indigenous peoples and lands are among the most maltreated and exploited, together with those brought to these shores from Africa to provide the power to drive colonial plantations and other green imperial projects. As we struggle for an energy transition away from fossil fuels, there are other forms of energy exploitation to avoid, along with the struggle to build structures that promote greater energy justice. This involves reflecting on relational and energy economies and the religio-cultural and theological narratives that intertwine with them. Here I will reflect on alternatives to toxic petro-authoritarian oligarchies to help amplify and imagine what seems impossible right now—namely, what a theology of regifting relations for the planet may look like.

RE/GIFTING THE LAND AFTER PETROEXTRACTIVISM

> Everything in the universe may be described in terms of energy. Galaxies, stars, molecules and atoms may be regarded as organizations of energy. Living organisms may be looked upon as engines which operate by means of energy derived directly or indirectly from the sun. The civilizations or cultures of mankind [sic], also, may be regarded as a form or organization of energy.[7]

Energy desires and energy economics are tied with how we might imagine the world, how deity manifests and is experienced. The various energy economies of history have influenced human ideas of the sacred, of divine power, and of how relations among humans and to the energy sources around them have changed over history. If we want to change energy relations toward less oppression and more energy justice, countering the pollution of earth systems and the destruction of human communities, we must imagine different ways of doing and distributing energy systems. In these efforts toward energy justice and energy equity, indigenous women have been central. Cara Daggett suggests that "the blossoming of alternative energy stories that counter patterns of domination by fuel, and that feature political innovations that more equitably and sustainably organize energy, will be as important as new energy tools, and perhaps more so."[8]

Energy transitions often go along with societal, cultural, and economic processes[9]—and thus also religious transformations. Indeed, energy can be seen as one way to describe the Divine Economy, if seen as relations. A first step would be to employ a reconstructed concept of the Divine Economy as

> transformative relations at the cosmic level: what pains, weakens, twists, destroys, denigrates the sacred cosmos and its inhabitants, and what might heal, strengthen, recreate and uplift "all our relations" to creator, redeemer, and sustainer. This cosmic economy is without what neoliberal economists call "externalities," without an outside that can be neglected or exploited.[10]

Struggles to transition away from oil and gas must also take into account existing energy inequities that make it likely that some areas in the world are stuck with aging and polluting fossil infrastructure and stranded assets, as well as lagging upgrades to less polluting and more efficient technology. But even transitions to renewables without addressing energy justice contain their own dangers. Thus, "many critical energy scholars worry that a renewable transition, without being led by democratic and just processes, could just as easily work to entrench existing power relations, while at the same time failing to produce truly sustainable communities."[11]

One of the characteristics of indigenous economic thought revolves around giving thanks and the gift. Many Western theologians have engaged this theme via colonial anthropology—for instance, Marcel Mauss, Jacques Derrida, and others. Mauss stressed in his classic 1925 study *The Gift* that Romantic notions of "native" gift economies reflect more the cultural context of the researcher than that of the people observed. He suggested that the European search for a pristine gift economy in fact revealed that "there has never existed, either in the past or in modern primitive societies, anything like a 'natural' economy."[12] Mauss's observations are based on anthropological records that reflect a Western account of indigenous thought, one that is often still cited as if the field of anthropology had remained unchanged since. James G. Carrier argues that there was more variety in both European and native societies but that few anthropologists have studied the gift in the modern West, and others are ignoring "commodity relationships in village societies."[13]

Jacques Derrida reflected on the gift as an ambivalent affair, as a *pharmakon*, meaning potentially either "remedy [or . . .] poison" for a disease.[14] The gift can enrich and impoverish, poison and heal, the communities it moves between.[15] Gift and *pharmakon* "can never be simply beneficial." I propose that Derrida's notion of the impossible gift is reminiscent of the vital ambivalence this dynamic retains—that we should not seek to remove ourselves from the constraints of reciprocity, but instead resist the chimera of a gift without reciprocity that often marks exploitative forms of corporate speak and relationships.

Divine grace is generally described as undeserved, unforceable, and without appropriate response in Christian traditions. There are good reasons to assert this, as it counteracts problematic tendencies in human relationality to the Divine, such as the tendency to bargain, compel, and attempt to force by various means divine support for one's own purposes. John Milbank reserves redemption and forgiveness as a divine "true gift." God remains only a giver and is never a recipient in a gift exchange. Thus, the God-given gift is a "transcendental category" in a way that structures theological discourse about creation, grace, incarnation, atonement, the church, and spirit,[16] all of which have been described as a "gift." This gifting and the related *methexis* as a "sharing of being and knowledge in the Divine"[17] flow only in one direction: from God to humans and from there to other humans, but never toward "him."

Such "sovereign" lack of the need for reciprocity is echoed in representations of British colonial logic.[18] There are echoes between theological and imperial language, the image of the sun, who gives light but does not receive anything back; God, the ungiven giver; and the sovereign who "gives and does not expect." The theological image of grace as a gift parallels imperial propaganda of the ever-present, all-powerful sovereign. This image of God builds upon a long but increasingly problematic tradition of casting God as a propertied, autocratic male owner and humanity as an impoverished, lacking, feminized recipient. We are in our nothingness before "him,"[19] and yet the metaphor is unstable, and nothingness is gendered, a kind of "feminine lack" in giving.[20] This phrasing repeats classic tropes of divine economy such as the commerce of the *conubium* and perpetuates a problematic gendering of the human-divine relationship where a masculinized, propertied divine bridegroom seeks to marry a feminized, unworthy whore-bride, giving "her" a new body.[21] Such unidirectional power needs not to ask for permission, but it is also not responsible to the masses of feudal serfs, peasants, workers, and the disinherited. And it matches all too closely the power claimed by the authoritarian billionaire class and their promotion of toxic petromasculinity as the very snake oil to cure our every ill. I ask therefore: in a context of creeping environmental destruction and tragically unavoidable severe climate chaos, is it tenable to claim that divine grace does not need a responsible human agent? Does a sense of creation as "gift" in this context not compel a rethinking of what responsible human relationships to the land might look like? This involves rethinking what Daggett calls the "geo-theology of energy,"[22] implanted by narratives that have been formed during modern extractivism and provided "scientific validation of the Protestant Ethic of maximizing work and minimizing waste."[23] Theopolitical micronarratives thus became integral to petroculture and other extractivist projects. Changing them will be crucial for facilitating transitions into a post-petroleum and less extractivist age. We need to reimagine gift as Eu-Charis (the good gift) in ways that indeed are freely given but oblige responsible reciprocity if the gift is to be honored and life to be valued. It is key to avoid falling into more problematic appropriations of gift theories that mark fossil fuels or rare metals as "gifts" of the earth or of the Divine, but rather to resist the transition from gift to sacrifice as lands are being marked for destruction and mining to supply metals for electric car batteries. The temptation to substitute one sacrifice for another continues as humanity is trying to reconcile the reality of energy entropy[24]—the fact that it continues to dissipate—with its ongoing hunger for energy. But let us first return to a sacrifice zone for petropolitics and see what learnings for an ethic of gift as reciprocity might be obtained from the case of resistance against pipelines in the U.S. Midwest.

LAND USE AND SOVEREIGNTY

The 2016 #NODAPL Standing Rock pipeline protest made visible a shift in the spiritual and ecological economies of tribal communities; a trans-tribal alliance fought back against the delivery system of petroculture, the "Black Snake" of petroleum, and the attached infrastructure that delivers it via indigenous lands—pipelines.[25] Standing Rock provided a convergence point for indigenous groups who had become increasingly organized and activist on land and water issues. This case illustrates how extractive petroculture continues to dissect and destroy the lands and their integrity, just as the people of the land have been denied to live in that reciprocity for what are now centuries.

Further afield, indigenous peoples throughout Canada have protested several of the points at which oil from the tar sands of Alberta connects to the Western Canadian Coast via the Coastal Gaslink Pipeline.[26] Like the protests at Standing Rock, the Canadian tribal group of the Wet'suwet'en resist the collapsing of their lands and livelihoods, while others demonstrate against the development of Line 3 in Minnesota.

I had the opportunity to spend a few days at Standing Rock, days that offered a glimpse that "another world is possible," one in which a different way of relating to the land, to the Sacred, and to each other was palpable as a possibility for entering a more conscious state of concrescence. Standing Rock provided "for a brief moment in time, a collective vision of what the future could be."[27] I have since been wanting to contribute to imagining these possibilities, including the many that are necessary. Water protectors encouraged us to take the movement home with us and to continue imagining its next steps: work locally to educate and engage more people, divest from petroleum dependency, and imagine alternatives.

Land rights issues reach back into the time of British coloniality, in this case a coloniality that is ongoing in the form of the Canadian government in which giving and taking have been deeply contested:

> The instability of colonial give and take is exemplified in accounts of "Indian giving" as leaving the status of the colonists undecided as taker, traders, or thieves, as a counterpoint to Europeans' long-standing presentation of themselves as bringers of gifts, specifically, the gifts of civilization and Christianity. . . . These were gifts so huge that they dwarfed any negative aspects that might come along with

them and justified taking the bounty of the New World. So powerful and persistent has been the idea of Western civilization as the source of all that is important and valuable that recent emphasis on what "we" owe to Indians culturally as well as materially is felt to be claiming something new and controversial.[28]

Stephen Greenblatt's lucid investigation of the theological underpinnings of such gift rhetoric suggests that giving and taking can be mobilized in multiple ways and by different actors.

The "New World" was considered to be possessed of "an abundance that was permanent and natural and therefore needed none of the hierarchies and hoardings that followed from a scarcity economics,"[29] although the lust for gold soon did away with such romantic ideas. Assumptions of divine and natural abundance merged with other constructions to form the dissonant, mixed messages of colonial imagery, "a persistent presentation of Europeans as the givers, and a stress on the power of giving, *alongside* the counter idea of the New World as a bountiful, fecund place from which to take."[30]

This resonance of the double givingness serves as an apology for empire as royal and divine sovereignty are merged to subordinate land and inhabitants while claiming a superabundance of giving and of resources that are inexhaustible. Under these ambivalent conditions, colonial encounters have caused culture clashes that find indigenous peoples struggling to find constructive ways to survive and thrive under colonial hegemony. The destruction of indigenous people and the earth have long been linked, and ecofeminists have long highlighted that women in particular bear the brunt of ecocidal colonial violence.

This violence toward women and land is especially visible in petroleum extraction. The "man camps" that arise as petro-infrastructure is built have long been connected to the rape, abduction, and vanishing of indigenous women. In North Dakota the incidents of violence and abuse against women rose sharply during the time of the Bakken oil boom, and close to 100 percent of this violence is committed by non-indigenous men, bringing new meaning to the term "toxic masculinity."[31]

After the #NODAPL camps were disbanded, activists went back to their home community to pursue action in various ways: by rebuilding indigenous communities, pushing for divestment from fossil industries, by studying, by creating rituals that help articulate grounding earth spiritualities, or by rethinking economies for a vulnerable planet. Rituals help engage new habits of thinking and give us new aspects of story and inspiration. The spiritual aspects and the rituality at Standing Rock invited people to find an inner sense of prayer—or commitment—and community that would help ground them for work in other locations. The camp attempted to instantiate instances of gift economy as people shared places to sleep, clothes, sleeping covers, food, instruction, and conversation, even for just a few days or weeks, invoking the energy of a potlatch.

GIFT, OBLIGATION, RECIPROCITY: OF POTLATCH AND PIPELINES

> Rather than to greed, prosperity here gave rise to the great potlatch tradition in which material goods were ritually given away, a direct reflection of the generosity of the land to the people. Wealth meant having enough to give away, social status elevated by generosity.[32]

In his investigation of European/North American discourses around native practices of potlach, Christopher Bracken contends that the Canadian legislation to forbid the potlatch and the controversy around it implied that "when Canada finally delivered itself to its western border, it found Europe already embodied in a group of cultures that white Canadians wished to define themselves against. Europe was already there among the very First Nations that European Canada, Europe-in-Canada, considered absolutely different from itself."[33] Bracken argues that the notion that there had once been societies that practiced a "true gift" was a quite recent invention: "For Mauss the distinction between a gift, conceived as an event that brings nothing back to the giver, and an exchange, understood as a reciprocal circulation of goods and services between two or more parties," is a "fairly recent" development, "peculiar to Western European societies." Instead, Mauss argues, there is no such thing as a free gift, and obligations and limitations are always part of exchanges, whether in "civilized" or "primitive" settings. Rather, he critiques one of the basic hermeneutic moves of early anthropologists and ethnographers, that the indigenous social system represented an unchanged, archaic, primitive system of social arrangements that was interesting perhaps primarily as the (imagined) "antecedent" of the cultures that Europe sees itself as representing, depending on one's position, an apex of civilization or a perversion of social accountability. Jacques Derrida writes the following on Mauss's influential book:

> Now this equivalence of giving-taking is precisely stated in the form of a "beautiful Maori proverb" that ... comes to close the "Moral Conclusions": "Ko Maru kai atu Ko Maru kai mai Ka ngohe ngohe." ... "Give as much as you take, all shall be very well."[34]

Reciprocity, respect, and balance here manifest as the moral that allows for the continuation of life and community, undermining the search for the impossible gift. Theologies seeking to decolonize Christian traditions and imagine a more equitable economic and social world can learn from rethinking economies of giving interculturally.

> Human understanding (as opposed to human control) requires reciprocal exchange, for all its hazards—your wisdom for mine (wanaanga atu, wanaanga mai), as we cross our thoughts together (whakawhitiwhiti whakaaro). In New Zealand, at least, a collaboration between Maori and Western knowledges seems possible. It may lead, eventually, to studies of cross-cultural encounters that do justice to the ancestors on both sides, and the potent, perilous pae—the edge between them.[35]

North American indigenous scientists like George Cajete and Robin Wall Kimmerer contribute to such cross-cultural encounters of gift and reciprocity. This also applies to post- or decolonial understandings of mission. The legacy of the abuse and perversion of gift reciprocity is part of the church also, and theologies that correct and reconstruct these theologies of Divine Economy must consider it as a reciprocal economy, not one where God is the "ungiven Giver" and humanity has nothing to offer in return, but rather a reciprocal mutuality.

Such impulses are also present in classical theologies that highlight the importance of doxology as the root of theology and liturgy and have wanted to develop more deeply the reciprocal potential in certain liturgies of the Eucharist. The Greek word "Eucharist" derives from good gift, *eu-charis*, that is shared around the altar and could also be rendered as thanksgiving. As a sacrament, the radical sharing economy hinted at in the Eucharist provides a ritual reminder of the greater sharing of the creation, the sharing of food for the hunger of all created being, providing the energy that is needed for bodies to function. We participate in each other's life and being, initiated into each other's pain and suffering as one body with many members.[36] Reciprocity is essential for a gift exchange to be more than exploitation or a romantic sentiment. Wall Kimmerer writes, "Gifts from the earth or from each other establish a particular relationship, an obligation of sorts to give, to receive, and to reciprocate."[37]

This, then, is an objection to the Derridean logic of the "gift" vs. economy that I never could quite myself articulate. Wall Kimmerer elaborates: "A gift is something for nothing, except that certain obligations are attached." Thus, the gift passes "from hand to hand, growing richer as it is honored in every exchange," because the "fundamental nature of gifts [is that] they move, and their value increases with their passage."[38] She then explains that the term "Indian giver" has been used negatively to describe somebody who "gives something and then wants to have it back" and is best understood as a "fascinating cross-cultural misinterpretation between an indigenous culture operating in a gift economy and a colonial culture predicated on the concept of private property."[39] Both sides thought the other was disingenuous, and as these two economic principles collided, indigenous people "understood the value of the gift in reciprocity and would be affronted if the gifts did not circulate back to them," while to the settlers the gift meant that it changed ownership and use-rights permanently:

> From the viewpoint of a private property economy, the "gift" is deemed to be "free" because we obtain it free of charge, at no cost. But in the gift economy, gifts are not free. The essence of the gift is that it creates a set of relationships. The currency of a gift economy is, at its root, reciprocity. In Western thinking, private land is understood to be a "bundle of rights," whereas in a gift economy property has a "bundle of responsibilities" attached.[40]

Such "bundles of responsibility" then are not to be bought or sold. Plants interact with us, and we are part of the reciprocity with them, where "they can't meet their responsibilities unless we meet ours," and through "reciprocity the gift is replenished" and flourishing is mutual.[41] Cultures of reciprocity go beyond cultures of gratitude, Wall Kimmerer distinguishes.

The call of the Water Protectors at Standing Rock—Mni Wiconi! "Water is life"—reminds that water is a gift that calls to be managed in responsibility. Instead, "water has been tricked"; it has been polluted, bottled, fracked, "corrupted, and instead of a bearer of life, it must now deliver poison."[42] Water comes with strings of responsibility attached, and disregard of those responsibilities results in destruction.

At Standing Rock, women performed the water ritual each morning, highlighting the responsibility of women in this spiritual economy of thanksgiving.[43] Women are particularly considered to be water protectors.[44] Youth, and in particular girls, were central to the organizing at the site and have been vocal in climate protests since. Defeating the Black Snake of the petroleum economy is both a spiritual and economic quest. Dismantling the toxic streams of oil and its spills, making it unprofitable, shifting investments, and so forth seem to finally show some success. Some observers have seen parallels between the young women at Standing Rock who led the fight against the Black Snake and Mary the Virgin, who is often depicted crushing the head of a snake.

Whatever imaginary folk may be braiding, if some part of us is to survive and enter into surely very different covenantal relationships, we have to relearn how to live in a changed world in cross-cultural covenants of reciprocity. That wisdom of life can be threaded from biblical narrative, as from indigenous creation stories. The breakdown of relationships of respect and mutuality marks many creation stories. The indigenous culture hero Skywoman Falling depends for her survival on muskrat, who offers its life as sacrifice to bring up the mud from which Skywoman's seeds can develop. Timothy Leduc's cross-cultural theological reflection on the lands and spiritual geography of Toronto and Eastern Canada shows Skywoman's figure as related to Notre Dame/ Our Lady, the Stella Maris, the star over the waters, each ancestral woman whose presence and wisdom has helped frame the spiritual worlds she represents.[45] Leduc suggests a braided recovery of the spiritual ancestors of all Canadians, bringing together indigenous and settler ancestors and their braided descendants and narratives in order to imagine a "climate of mind" that can locate ancestral and present sacrality in land and urban environment. Wall Kimmerer envisions similarly a "polyculture of complementary knowledges" that is necessary to defeat monocultures of science and production.[46]

Nanabozho, the culture hero of Anishinaabe culture, represents the personification of life force and its duality. He had a twin brother committed to making as much imbalance as he was dedicated to engendering balance—a brother whose arrogance of power pushed unlimited growth and threw humility into the wind,[47] reflecting the duality of human nature that is prevalent in the narratives of many cultures. Further, the Anishinaabe monster Windigo represents that within us which cares more for our own survival than anything else, with a heart made of ice and a ravenous hunger that increases the more it devours, ending in uncontrolled consumption.[48]

The blessing/curse scenarios of the Deuteronomic tradition seem to illustrate a similar dynamic: respect good relations, or you will be disrespected, and the land will reject you. Ironically, it is this emphasis on reciprocity that Christians have often read as oppressive, as the jealousy of the Old Testament God, a favorite supersessionist reading strategy.

Recent studies have now also more closely explored the ideas of gift in biblical texts. John Barclay reminds us that Mauss suggests that the idea of a gift given without concern for reciprocity is a modern invention—in particular, that of Reformation theologians.[49] More specifically, he argues that in Graeco-Roman and Jewish practices "a gift can be *unconditioned* (free of prior conditions regarding the recipient) without also being *unconditional* (free of expectations that the recipient will offer some 'return')."[50] Indeed, "benefits were generally intended to foster mutuality, by creating or maintaining social bonds," where the expectation of reciprocity "created cyclical patterns of gift-and-return" and thus, what seems in modern terms to be paradox—"that a 'free' gift could also be obliging—is entirely comprehensible in ancient terms."[51] The fact that a benefit is undeserved and cannot be earned does not mean it does not oblige. Further, it is important to note that in all this, Paul "stands *among* fellow Jews in this discussion of divine grace, not apart from them in a unique or antithetical position."[52] This counteracts the supersessionist readings of Pauline grace as radically innovative or different from all Jewish traditions.

Perhaps what is also visible is an echo of "cultures of gratitude," an economy of thanksgiving that is present also in the biblical imaginary but that in the Reformation has been superseded by an economy of (cheap) grace. Has this grace, especially in its iterations as Protestant work ethic, lured us away from these relations of gratitude and reciprocity, or can rereadings of grace as the undeserved divine gift that invites reciprocity be heard anew? Cultures of gratitude necessitate lives of respect and reciprocity rather than mere lip service. Catastrophic climate change endangers cultures of reciprocity and gratitude; therefore, it is important to develop resources for people to imagine different ways of being. A political spirituality involving an eco-justice approach to energy will need to be part of this effort.

The basic democracy that is at the heart of a more respectful relationship between land, water, and its creatures has long been threatened by the autocratic, exploitative hierarchies of those that seek to drain the last drop of wealth from the soil and waters. Rachel Maddow, following initially a more political track of corruption and anti-democratic maneuvering of white supremacist billionaires on either side of the old Cold War lines, found that one of the major ties that binds oligarchs and billionaires with autocratic political leanings together is anointing with oil.[53] Defeating the Black Snake is both a spiritual and economic quest. Dismantling the toxic streams of oil and its spills, making it unprofitable, shifting investments, and so forth seem to finally show some success. And yet, some pipeline projects persist, coupled with a hell-bent aggressive form of climate change denial, silencing, and counter-messaging of scientists and protestors that expresses petromasculinity in a caricature of gender essentialism often directed at young women, indigenous peoples, and others who engage themselves in climate change action or just happen to live near extraction sites where they become victims of the toxic desires of oil-consuming economies. Drilling into the flanks of the earth with petro-toxicity manifests as tragic, parallel with seeking to penetrate young, vulnerable women, just as ecofeminists have described for decades. The rape of earth and women is not just metaphorical; it is in our faces, as the earth is being diminished as her fever spikes each day.

TOWARD ENERGETIC TRANSITIONS

A future that can enable survival of the species of the creation community necessitates a different theopolitics of energy, reconceiving "energopower,"[54] or "geo-theology."[55] Ultimately, we need a different way of conceiving, producing, and using energy. Better energo-theologies will help us understand anew how the Divine manifests in this time of rapid climate change and societal transformation.

NOTES

1. Robin Wall Kimmerer, *Braiding Sweetgrass: Indigenous Wisdom, Scientific Knowledge, and the Teachings of Plants* (Minneapolis: Milkweed, 2013), 32.
2. A term used by Sarah Kendzior, cohost of *Gaslit Nation* podcast: https://gaslitnation.libsyn.com/kleptocracy-world-order.
3. Marion Grau, *Of Divine Economy: Refinancing Redemption* (London and New York: T&T Clark/Continuum, 2004), 173, quoting Lewis Hyde, *Trickster Makes This World: Mischief, Myth, and Art* (New York: Farrar, Straus and Giroux, 1998), 11.
4. The phenomenon of the "confidence man" is well known in U.S. popular culture and documented among other places through Herman Melville's moral parable *The Confidence Man: His Masquerade*, published on April Fool's Day, 1857.
5. Nancy MacLean, *Democracy in Chains: The Deep History of the Radical Right's Stealth Plan for America* (London: Scribe, 2017), 1.
6. See, for example, Timothy Mitchell, *Carbon Democracy: Political Power in the Age of Oil* (New York: Verso, 2011); Rachel Maddow, *Blowout: Corrupted Democracy, Rogue State Russia, and the Richest, Most Destructive Industry on Earth* (New York: Crown, 2019); and Darren Dochuk, *Anointed with Oil: How Christianity and Crude Made Modern America* (New York: Basic Books, 2019).
7. Leslie White, "Energy and the Evolution of Culture," *American Anthropologist* 45, no. 3 (1943): 335.
8. Cara Daggett, "Energy and Domination: Contesting the Fossil Myth of Fuel Expansion," *Environmental Politics* 30, no. 4 (2021): 644–62.
9. Thomas Schlabbach and Viktor Wesselak, *Energie: Den Erneuerbaren Gehört die Zukunft* (Berlin: Springer, 2020), 1.
10. Marion Grau, "From Burning Bush to Refiner's Fire: Reflections on the Combustive Element of Fire," in *The Bloomsbury Handbook of Religion and Nature*, ed. Laura Hobgood and Whitney Bauman (London: Bloomsbury, 2018), 161.
11. Daggett, "Energy and Domination," 14.
12. However, Mauss's own fictions include that he was able to describe "the gift" in a way that has tended to universalize that notion across differences and hence potentially propagate "orthodox" notions about "the gift." Thus, his account has at least two effects: a reality check for Western notions of a "pure gift" and a tendency to claim a total description; Marcel Mauss, *The Gift: Forms and Functions of Exchange in Archaic Societies* (New York and London: W. W. Norton, 1967), 3.
13. James G. Carrier, "Maussian Occidentalism," in *Occidentalism: Images of the West*, ed. James G. Carrier (Oxford: Clarendon Press, 1995), 95.

14. Note that the German word *Gift* is rendered as "poison" in English; Jacques Derrida, *Dissemination*, trans. Barbara Johnson (Chicago: University of Chicago Press, 1981), 94. Hence one could say in Germlish *Das Gift vergiftet* (The gift poisons); see John D. Caputo, *The Prayers and Tears of Jacques Derrida: Religion without Religion* (Bloomington and Indianapolis: Indiana University Press, 1997), 166.
15. For an exploration of the various ways in which gifts can function, see Derrida's discussion of "Plato's Pharmacy" and especially some of the sections on the *pharmakon* in Derrida, *Dissemination*, 95–155.
16. John Milbank, *Being Reconciled: Ontology and Pardon* (London and New York: Routledge, 2003), ix.
17. Milbank, *Being Reconciled*, ix.
18. David Murray, *Indian Giving: Economies of Power in Indian-White Exchanges* (Amherst: University of Massachusetts Press, 2000), 55.
19. Note Milbank's insistent use of the male pronoun for God; Milbank, *Being Reconciled*, 46 et passim.
20. I have explored the issue of the feminization of giving in strands of traditional interpretations of women in the gospels and their excessive giving (which has also been read as foreshadowing Jesus's abundant giving of his life) in more detail in Grau, *Of Divine Economy*, 99–107.
21. John Milbank, "The Midwinter Sacrifice," in *The Blackwell Companion to Postmodern Theology*, ed. Graham Ward (Oxford: Blackwell, 2001), 128.
22. Cara Daggett, "Energy and Domination," in *The Birth of Energy: Fossil Fuels, Thermodynamics & the Politics of Words* (Durham, N.C.: Duke University Press, 2019).
23. Daggett, *Birth of Energy*, 54.
24. Scholars of energy and energy humanities point out the constraints that the fluctuations of energy described by the Laws of Thermodynamics put on any human attempt to counter these fluctuations. Some describe entropy as "depressing" (Braun and Glidden) or otherwise as having caused theological crises (Daggett), since entropy was understood as undermining the sense that humanity stood at the center of creation; see Timothy F. Braun and Lisa Glidden, *Understanding Energy and Energy Policy* (London: Zed Books, 2014), 15; and Daggett, *Birth of Energy*, 59.
25. Marion Grau, "The Camp Is a Ceremony: A Report from Standing Rock," *Religion Dispatches*, https://religiondispatches.org/decolonizing-thanksgiving-at-standing-rock-a-black-friday-report/.
26. *Unist'ot'en*, http://unistoten.camp/. Canadian church leaders have released a letter voicing support of indigenous resistance against this pipeline project:

"Church Leaders Sign Statement of Support for Wet'Suwet'en," *Anglican Journal*, https://www.anglicanjournal.com/church-leaders-sign-statement-of-support-for-wetsuweten/.

27. Nick Estes and Jaskiran Dhillon, "The Black Snake, #NODAPL, and the Rise of a People's Movement," in *Standing with Standing Rock: Voices from the #NODAPL Movement* (Minneapolis: University of Minnesota Press, 2019), 5.
28. Murray, *Indian Giving*, 18.
29. Murray, *Indian Giving*, 51.
30. Murray, *Indian Giving*, 60; my italics.
31. Michelle Cook, "Striking at the Heart of Capital: International Financial Institutions and Indigenous Peoples' Human Rights," in *Standing with Standing Rock*, 112–13.
32. Kimmerer, *Braiding Sweetgrass*, 280.
33. Christopher Bracken, *The Potlatch Papers* (Chicago: University of Chicago Press, 1997), 2. Hence, it is no surprise that, as Bracken further contends, Mauss's investigations were conditioned not so much by his Romanticism about "primitive peoples," but by his culture's unconscious (155).
34. Jacques Derrida, *Given Time*, vol. 1, *Counterfeit Money*, trans. Peggy Kamuf (Chicago: University of Chicago Press, 1992), 67.
35. Anne Salmond, *Between Worlds: Early Exchanges between Maori and Europeans 1773–1815* (Honolulu: University of Hawai'i Press, 1997), 513.
36. William T. Cavanaugh, *Being Consumed: Economics and Christian Desire* (Grand Rapids, Mich.: Eerdmanns, 2008), chapter 4.
37. Kimmerer, *Braiding Sweetgrass*, 25.
38. Kimmerer, *Braiding Sweetgrass*, 27.
39. Kimmerer, *Braiding Sweetgrass*, 27–28.
40. Kimmerer, *Braiding Sweetgrass*, 28.
41. Kimmerer, *Braiding Sweetgrass*, 140, 166.
42. Kimmerer, *Braiding Sweetgrass*, 315.
43. Kimmerer, *Braiding Sweetgrass*, 94.
44. Kaitlin Curtice, *Native: Identity, Belonging, and Rediscovering God* (Grand Rapids, Mich.: Brazos, 2020), 10.
45. Timothy B. Leduc, *A Canadian Climate of Mind: Passages from Fur to Energy and Beyond* (Montreal: McGill-Queen's University, 2016), 156.
46. Kimmerer, *Braiding Sweetgrass*, 139.
47. Kimmerer, *Braiding Sweetgrass*, 205, 212.
48. Kimmerer, *Braiding Sweetgrass*, 304–5.
49. John M. G. Barclay, *Paul and the Gift* (Grand Rapids, Mich.: Eerdmanns, 2015), 37.
50. Barclay, *Paul and the Gift*, 312.

51. Barclay, *Paul and the Gift*, 312.
52. Barclay, *Paul and the Gift*, 314.
53. Maddow, *Blowout*.
54. Dominic Boyer, "Energopower: An Introduction," in *Energy Humanities: An Anthology*, ed. Imre Szeman and Dominic Boyer (Baltimore: Johns Hopkins University Press, 2017), 184–205.
55. Daggett, *Birth of Energy*, 51ff.

❖ The Immanence and Transcendence of Christianity, Capitalism, and Economic Democracy: Alternatives to Ecological Devastation

JOERG RIEGER

INTRODUCTION

Western Christianity has often been suspected of being a major contributor to the rampant ecological devastation that marks our age, if not its cause.[1] While some see the problem with Christianity's anthropocentrism, playing off humanity and the earth and leading to the devaluation and destruction of the latter on a grand scale,[2] others find fault with Western Christianity's emphasis on the nonmaterial, ethereal, and transcendent. In response to the latter problem, scholars of religion and theology have sought to give more prominent voice to religious traditions that emphasize the material, non-ethereal, and immanent. Yet while religions that value the material and immanent over the immaterial and transcendent might seem to be poised to promote healthier concerns for the environment, this is not necessarily the case.

The so-called Gospel of Prosperity may serve as one example. Its celebration of wealth at the top—whether the wealth of its pastors or other prominent members of the community—not only lacks ecological concern but does little to alleviate the burdens imposed on the environment. The private jets of its most notorious representatives symbolize an interest in the material and immanent that is not only ecologically unsustainable but also never in reach of the masses. Seeking to imitate this kind of economic success, most of the followers of the Gospel of Prosperity fail while further damaging both their communities and the environment.

Capitalism, of course, serves as the primary example of how dominant concerns for material wealth not only fail to protect the environment but are directly implicated in its exploitation and destruction. In this context, putting greater value on material things—for instance, through well-meaning efforts to shift the natural environment from an externality to a more integral part of capitalist economic calculations—may bring some relief, but it does not change the fact that the main goal of capitalist economics remains turning (for the few) a profit rather than saving the world (for the many).

RELIGION, IMMANENCE, AND THE MATERIAL

For those who find themselves drowning in oceans of religious, political, academic, and sometimes even economic idealism, engaging the value of immanence and material reality provides some relief. Such engagements need not waste time with the crude materialisms that rule the day in certain discourses of the natural sciences or with now mostly discredited reductionist economic determinisms. Reductionist accounts are of little help in developing the bigger picture we need.

Against the backdrop of materialist reductionism, the current discussions emerging in the so-called new materialisms appear to be more fruitful. At stake is no longer the abortive discussion of whether "material" or "ideal" factors are all-determinative; the question is not even which factors are primary and secondary, but how these factors influence and shape each other. In this way, new materialisms reclaim aspects of the older dialectical materialisms. Moreover, new materialisms can contribute to a constructive rethinking of religion, how it intersects with and is part of material reality, and what difference it makes. In addition, new materialisms also invite analyses of the existence of alternative religious practices that do not conform to the dominant powers, providing deeper investigations of their nature and promise.

New materialist scholars of religion invite the fields of religious studies and theology to take more seriously material and physical realities and to reconceive the roles of ecology, energy, and even economics in the production of religious experience.[3] While this increased emphasis on the material does not mean that religion necessarily works for the good, in these accounts religion can be examined for its potential to become a force for empowerment and social change. Such broadening of older dialectical materialist traditions is promising for various reasons.

First, material reevaluations of religion resonate with many of the Jewish traditions that shaped Christianity. It has often been observed that salvation in the ancient Hebrew traditions is not about going to heaven after death but about the flourishing of life in the present. God's promise to Moses and the Hebrew slaves is not that their souls will live forever in heaven but that they will be liberated from slavery and will be led into a land flowing with milk and honey (Ex 3:7–10). What nineteenth-century European interpreters rejected as the primitive spirit of Judaism in comparison with Christianity—focusing on the concerns of the immanent rather than the transcendent—is now being reclaimed and revalued.

Second, reevaluating religion in immanent and material terms opens the view for a wealth of new insights produced by the natural sciences, from quantum physics and genetics to neurobiology. The natural sciences have come a long way from the days of Newtonian physics, where cause and effect, subject and object were easily distinguished and every question had a straightforward answer. As the sciences are taken more seriously in the humanities and religious studies, we need to keep in mind that "sciences (and technologies) and their societies co-constitute each other," as Sandra Harding has pointed out from a feminist and postcolonial perspective.[4] In other words, this new focus on immanence is complex, composed of various embodiments of immanence—natural and social ones—including the "transcendent" superstructures of both science and society.

Third, new materialisms reshape and broaden our understanding of human agency, which does not need to be understood as originating in transcendent ideas or good intentions. Noting a "mismatch between actions, intentions, and consequences," new materialists Diana Coole and Samantha Frost advocate an open systems approach when considering interactions between socioeconomic and environmental conditions, combining biological, physiological, and physical processes. By the same token, matter can be seen as having agency in its own right, as new materialists emphasize "the productivity and resilience of matter."[5] Matter is, therefore, always in a process of becoming rather than merely being. A great advantage of these approaches is that the possibility of transformation does not depend on the intentions of well-meaning individuals, and neither does it depend on the ideas of philosophers, theologians, or preachers.

New materialism thus offers some inspiration for rethinking both the immanent and the transcendent, although we need to gain further clarity about who the agents are and how alternative agency can emerge under the conditions of neoliberal capitalism, which seeks to harness every form of agency for its own purposes, human as well as nonhuman. A broadened analysis of capitalism is part of the conversation, as (in the words of new materialists Coole and Frost) "the capitalist system is not understood in any narrowly economistic way but rather is treated as a detotalized totality that includes a multitude of interconnected phenomena and processes."[6] Just as historical and dialectical materialisms emerged in the midst of the earlier tensions of capitalism, new materialism is at its best where it continues to engage these tensions today. This brings us back to the overarching question that the social and especially the ecological challenges of our age pose to us: "What are we up against?"

One approach to this question might be Coole and Frost's emphasis—not as dated as it might appear at first sight—on the "immense and immediate material hardship for real individuals" who lost their savings, their pensions, their houses, and their jobs in the meltdown of the economy after the Great Recession of 2007–9.[7] Over a decade later, most of the jobs that were created in the wake of the recession are not of the same quality as the jobs that were lost, with fewer benefits, lower compensation, and reduced influence and power at work. Reevaluating increasing environmental destruction and spiraling climate change in this light puts a new focus on conversations about ecology, as the most fundamental challenges might be neither anthropocentrism nor an overemphasis of transcendence but the structures of neoliberal capitalism. Reclaiming immanence without qualification (as in much of liberal theology), therefore, is not necessarily a solution but might be part of the problem.

REGROUNDING THE IMMANENT AND THE MATERIAL

Basic analyses of the economy and class tensions can help deepen conversations about the immanent and the material, although even some critics of capitalism seem to have lost sight of the issue. Current responses to financial capitalism, especially by theologians, have argued that all that matters in neoliberal capitalism is money and finance and that the market is now the transcendent—God.[8] While these arguments capture some aspects of what is going on, they overlook that material production still plays a significant role in capitalism's ultimate goal of increasing wealth. Why else would capitalism keep devouring both natural and human resources in more and more dramatic fashion? As Marx famously put it, labor is the father and nature the mother of wealth. Capitalism cannot increase wealth (the wealth of the shareholders, to be sure) without increasing production, even if production is not always of material things, like the production of services and software.

What is at the heart of the ecological devastations of our age is, therefore, both immanent and transcendent, material and immaterial, and all is shaped by the dynamics of capitalism, which is why it would make more sense to talk about the Capitalocene than the Anthropocene.[9] While religion is a strong contender in this story as well, religion in its most damaging forms—both for people and the earth—has been aggressively reshaped by the interests of big money, particularly in the more recent history of the United States since the 1950s.[10] To be sure, there are many historical precedents, like early Christianity being pulled ever more closely into the orbit of the Roman Empire. What matters for the development of our argument, however, is not generic assertions that empires by definition seek to control everything, including religion. What matters is the particular history of capitalism and religion and its implications for particular forms of exploitation that now reach to the ends of the earth and beyond.

While new materialists tend to operate with a stronger sense of the impacts of capitalism than scholars of religion, a more substantial assessment of neoliberal capitalist reality is needed. The historical analyses of older dialectical materialisms, for instance in various Marxist traditions, point the way by paying attention to the systematic exploitation of working people. The world of labor and production (both human and nonhuman) is where capitalism is rooted and where its greatest tensions manifest themselves, even in the days of financial capitalism. This is why the pushback against working people and their associations has been getting progressively worse, paralleled by the exploitation of the environment.

If capitalism is thus built on the productive and reproductive labor of people and the planet, the world of production deserves another look, not just in the study of economics but also in the study of everything else, including matters of the environment and religion. Note that production is not only the place of exploitation; it is also the place of resistance—this is another insight historically developed in materialist reflection. Productive and reproductive labor is where matters of agency and even the ability to think and believe differently are located. This observation is at the core of my argument.

Surprisingly, even new materialisms, not unlike the study of religion and theology, hardly mention labor or work. When labor is mentioned at all, as it is in one collection of essays that represents the spectrum of new materialisms, it is merely to make the point that new materialisms should go beyond the focus on labor that has been characteristic of materialisms in the past. Rosi Braidotti, a leading materialist feminist, develops her proposal as if labor did not even exist, proposing instead a "biocentered egalitarianism" that "breaks the expectation of mutual reciprocity," concluding that we have to give up ideas of retaliation and compensation.[11] While retaliation and "tit-for-tat" may indeed not be the most productive ways of relating to others, giving up notions of compensation and reciprocity in the current situation cannot be an option for people who have to work for a living—the proverbial 99 percent.

To move forward, new materialisms may need to take a step back. Where Crockett and Robbins "posit earth as subject" (including "materiality, energy forces, layered strata, atmosphere and magnetosphere, enfolded forms of life"),[12] what if we were to add working people as subjects? After all, working people are the agents who are doing most of the work that sustains humanity at present—think not only production and agriculture but also services and care, including both productive and reproductive labor—yet their contributions not only to the material but also to the immaterial (immanence but also transcendence) are often overlooked. A historically informed materialism cannot do without those whose subjectivity is neglected in dominant discourse, stretching from politics to economics and religion, where the assumption still rules that Caesar built the Roman Empire, Henry Ford produced automobiles, and Paul singlehandedly expanded the church's mission. How would the perspective of working people reshape reflections on the earth as agent or subject? How to reclaim these contributions, and how to link up with their revolutionary potential?

Regarding subjectivity and agency, the parallels are worth noting between that which devalues the agency of working people (by cutting salaries, benefits, hours at work, pushing employment at will, the gig economy) and that which devalues the agency of the nonhuman. Still, a difference remains in which the capitalist exploitation of labor and nature presupposes a hierarchy of labor over nature, reversing the ancient primacy of the productivity of land to the productivity of labor.[13] In this framework, unpaid reproductive labor, typically performed by women in patriarchal societies, is equated with nature, and the same is true for the unpaid labor of slaves. However, while unpaid reproductive human labor and natural labor are pushed lower than productive labor, it is the *sine qua non* of production, as the productivity of labor absolutely depends on it and therefore deserves special consideration.

Capitalism's efforts to downplay both productive and reproductive labor are mirrored even in the histories of religion and theology. The agency of working people is still mostly relegated to a footnote, even in many theologies that consider themselves liberative. And even where some of this agency is studied, this is done without understanding its importance for the development of religion as a whole. The study of popular religion, for instance, tends to study popular religious experience in itself, in isolation from dominant religious experience. In Christian theology, to give another example, Jesus preaching good news to the poor and uplifting them is noted often without linking it to the emerging agency of the people that pose a challenge to the powers that be. In Jesus's imagination, by contrast, even the stones can be agents in resistance against the forces of empire (Lk 19:40).

Karl Marx's critique of Ludwig Feuerbach's materialism highlights the agency of productive labor. Going beyond Feuerbach, Marx observes that material objects and matter itself are not mere givens but are produced by labor and commerce, which means that materialism needs to take into account the produced nature of matter. Material reality, in other words, is never a static entity, as it is constantly produced and reproduced, explicitly including the importance of reproductive labor performed by women, enslaved people,[14] and nature, which is still generally taken for granted. Even the so-called "natural" world is no longer "pristine" but has been shaped by human labor in some form or fashion. This fact did not escape Marx, even though the human impact on nature was less pronounced in his time than today, as the European landscape by then had already been substantially transformed by the cutting down of forests, the mining of coal, and various agricultural techniques. Human agency is inextricably tied up with the agency of nature, specifically in the agency of working people, as both nature and workers are constantly engaged in "changing the form of matter." This is where capitalism is rooted, as wealth is generated from the interplay of labor and nature. In the words of Marx, "[Humans] can work only as Nature does, that is by changing the form of matter. Nay more, in this work of changing the form he [or she] is constantly helped by natural forces."[15]

While the new materialisms have broadened our understanding of the productive capacities of nature far beyond what Marx and his contemporaries could have imagined, the interaction of labor and nature needs further investigation. In the current economic situation, a deeper analysis of the fundamental—and therefore potentially revolutionary—contributions of human labor (both productive and reproductive) might guide the way.

ECONOMIC DEMOCRACY RECLAIMING IMMANENCE AND TRANSCENDENCE

As we deepen these materialist intuitions, the tasks of the study of religion and theology are being redefined. Instead of studying disembodied ideas and seemingly universal truths, the task is now to study material relations and the respective ideas that are produced and reproduced in the context of particular relations of power. Let's call this the labor of religion, which develops in relation to productive and reproductive labor performed by working people, who make up the majority of humanity.

At stake is more than a methodological issue. Scholars of religion and theology will not be able to recognize alternative forms of religion unless they take into account the history of how power is shaped and reshaped in particular material social relations and social movements. As a result, we need to pay closer attention to whose agency matters in what ways and who benefits and who does not in a given system—keeping in mind the agency of both people and nature. This question of agency is at the core of all democratic traditions: who is calling the shots, who is in charge, who embodies power? These questions push beyond political democracy.

Reclaiming materialist traditions allows us to broaden our notions of democracy to include economic and labor relations on the one hand and religion and other cultural productions on the other. Why should the rule of people be limited to the political realm, whether defined as party politics or as broader matters of public interest? Why should places of work, where people spend the biggest block of time each week, be exempt from the principles of democracy? Why should people's productive activities at work be discounted from influencing the social and ecological changes we desperately need? And why not include religion and culture in conversations about democracy?

This, I would argue, is the fundamental challenge economic democracy addresses: how do we harness alternative powers that are already at work on the ground but commonly overlooked even by those who seek to organize and reorganize political democracy? Moreover, how do we push beyond elitist notions of democracy that we have inherited from Western traditions, according to which the primary agents are those who have the leisure and the time to engage in political activities? While, in contrast to the Greek polis and early forms of U.S. democracy that were more openly run by the elites (only 6 or 7 percent of the population even had the right to vote), the majority of citizens are now allowed to vote; the actual political agency of most people is still fairly limited.

Economic democracy includes the nonhuman, but in a less abstract sense than in many new materialist discourses. First, taking a closer look at productive and reproductive labor will help us become more aware of the produced nature of everything that surrounds us, including politics, nature, and religious practice. Discovering the produced nature of everything does not imply a negative judgment, as if being produced does not mean of secondary importance; it simply reminds us of the fact that nothing ever just "fell from the sky," not even religion or nature, and—this is the most important point to my argument—that there may be alternative modes of production that can be explored and harnessed.

Second, a closer look at production reveals its complexity, as production is both material and immaterial, with implications for both immanence and transcendence. There is, for instance, an odd sort of transcendence that occurs when produced objects are commodified. Marx uses the example of a table. In terms of its use value, there is nothing mysterious about a table. Wood, produced by nature, is transformed by human labor into the production of a common thing to be used for particular purposes: a dining-room table, a desk, a kitchen table. A certain kind of transcendence enters in terms of the exchange value of the table. In the economic exchanges of capitalism, the table becomes a commodity, and what matters is no longer the labor, the materials, or even its use value, but the profit that can be made when this table is sold. Because profit is usually thought of as a relationship between things, what is concealed is that it is actually produced in a relationship between people.[16] In Marx's own words, under the conditions of capitalism "a commodity is therefore a mysterious thing, simply because in it the social character of men's labour appears to them as an objective character stamped upon the product of that labour; because the relation of the producers to the sum total of their own labour is presented to them as a social relation, existing not between themselves, but between the products of their labour."[17] Marx compares this to religious ideas, where "the productions of the human brain appear as independent beings endowed with life."[18] What happens when the material and immaterial are made visible in these seemingly mundane processes of everyday labor, and what alternatives might emerge?

Keep in mind that while the significance of labor and of human relationships is concealed by the capitalist process of commodification (which then leads to commodity fetishism), labor and human relationships remain fundamental. The new materialist Sara Ahmed adds an important insight that extends production and labor to reproductive labor: the example of the table points to other divisions of labor, manifest for instance in the division between who usually works at desks and who usually works at kitchen tables. In this example, the kitchen table represents the racial, gender, and class-based divisions of labor, as desk work, for example, is often supported by the domestic labor of black and working-class women.[19] Moreover, reproductive labor (of which the kitchen table is only one example) is often valued more like the labor of nature, as a so-called externality that is either taken for granted and unpaid or pushed to the very bottom of the ranks of agency. Literary scholar Daniel Hartley has highlighted the important role that culture plays here: "If women's work has been historically vital to capitalism, then we must conclude . . . that culture is more than a force of ideological *legitimation*, it is itself a materially *constitutive* and *productive* moment in capitalist value relations."[20] In other words, deeper investigation of work does not need to sacrifice race, gender, and sexuality to class; and it reminds us of the importance of these matters to ecology!

Observing labor relationships matters not only for analytical purposes but also for a determination of what kind of alternatives might emerge from specific labor situations, both material and immaterial, immanent and transcendent. Economic democracy—embodied where most people spend the bulk of their waking hours and where most of their sustenance is rooted—has a tremendous impact on political democracy. That so much of politics is determined either by people who enjoy the privilege of being agents at work or simply by economic prowess (i.e., large amounts of money) is directly linked to the lack of economic democracy. And with regard to religion, not only will the images of God developed by minority women of the working class be different; they also provide much-needed challenges to dominant images of God.

Sociologist Jason Moore concisely formulates the task before us when he argues that "the organization of work—inside and outside the cash nexus, in all its gendered, semicolonial, and racialized forms—must be at the center of our explanations, and our politics."[21] Jason Edwards, drawing the conclusions of a collection of essays on the new materialisms, puts economics and politics together when he argues for a return "to a kind of historical materialism that focuses on the reproduction of capitalist societies and the system of states, both in everyday practices of production and consumption and in the ideological and coercive power of states and the international system."[22] At the core of it all is a relational view of power that emphasizes labor relations as an essential part, with implications for the study of religion and theology as well as ecological and environmental concerns. Note that this parallels my repeated interventions regarding the study of class as relationship rather than stratification.[23]

RESHAPING TRANSCENDENCE AND IMMANENCE AND THE STUDY OF RELIGION, THEOLOGY, AND ECOLOGY

The reflections so far can help reframe the options in the study of religion and theology. What Marx calls religion in the previous example is how philosophical idealists and many theologians view religion: as (seemingly) "independent things endowed with life." But there is no reason religion cannot be defined differently from a more materialist perspective that takes into account the relation of life to material realities. The same is true for the notion of transcendence: why define it in reference to the world of ethereal ideas, the mind, or whatever is beyond nature or otherworldly? Likewise, transcendence does not have to be determined by capitalism, either, in the transcendence of the commodity that Marx exposes previously or the flow of money in financial capitalism.

There are materialist ways to conceive of transcendence—for instance, when it is defined not in opposition to immanence but as transcending one kind of immanence in favor of another. This corresponds with the Jewish traditions, where salvation is not about escaping to another world but about the flourishing of life here and now, as noted earlier. In Christianity, this is one way to understand the incarnation of Christ, where the

Roman Empire is transcended not by ethereal ideas or otherworldly dreams proclaimed in sermons but by God's solidarity with peasant movements with whom the construction worker Jesus of Nazareth stands in solidarity. In this case, transcendence is what interrupts the status quo (an idea also expressed by the Jewish philosopher Emanuel Levinas in his *Totality and Infinity*), and that which is diametrically opposed to the status quo—totally other.

Such notions of transcendence are not as foreign to theology as they might appear. The quintessential modern theologian of transcendence, Karl Barth, whose basic conversion from liberal theology to a more radical approach took place in close relation to the labor movement in Switzerland, realized that to confess "God in the highest" does not mean to look up to the sky or away to other worlds. Recalling the birth of Christ in a stable in Bethlehem, Barth notes that this is the proper place of transcendence: "The highness of God consists in His thus descending."[24] This is the point of talking about the divine as the wholly other—*der ganz Andere*. In contrast to the liberal-traditionalist tug-of-war between immanence and transcendence that still shapes so much of theology even today, both the concepts of transcendence and of immanence are transformed here. Transcendence, we might conclude, is that which totally reshapes dominant immanence.

While capitalism needs to cover up the immanent contributions of productive and reproductive labor (both human and nonhuman), the study of religion can resist this coverup by taking labor (both human and nonhuman) into account, thus reshaping whatever might be defined as its transcendent or transcending tasks. At the core of my argument here is one of the fundamental shifts that I have suggested during the Great Recession: the move from a focus on the widespread concern for the redistribution of wealth to the production of wealth.[25] While the concern for the redistribution of wealth remains important in a time of rising inequality, it tends to overlook the roots of inequality, which are located in how productive and reproductive labor are valued. If those who are engaged in production and reproduction were appreciated appropriately (rather than pushed out of sight and exploited, like in capitalism), the distribution of wealth would be corrected as well. Much of this is true also for the environmental injustices that we see today, from environmental racism to food injustice and the utter exploitation of natural resources: the challenge is not just inequality but the place of the productive and reproductive labor of nature and the environment itself.

In conjunction with developing new appreciation for human production and reproduction and its revolutionary potential, ecological and environmental production and reproduction can also be seen in a new light. Rethinking working people as subjects in this way will illuminate notions of the earth as subject. Whitney Bauman's notion of "biohistories" could be helpful as we reevaluate the contributions of the planet and its agency.[26] Note that this conversation pushes beyond the typical efforts of gaining more respect for the environment or accounting for the environment in economic calculations; it is about becoming aware of and supporting new forms of the agency of nature that not only resist ecological devastation but bring forth new kinds of ecological flourishing that benefit both humans and nonhumans. The immanence that emerges here has transcendent qualities, as it fundamentally challenges and reshapes the immanence of the status quo.

Theologians might think of images of Gaia and proclamations of the earth as God's body (Sallie McFague), but it might be more helpful to think about earth's agency in more specific terms, related to productive and reproductive labor. This includes the production and reproduction of life at all levels, including the work of plants and animals, the majority of which are now domesticated. Wild animals, for instance, account only for 3 percent of the total mass (in weight), humans account for 30 percent, and domesticated animals for 67 percent.[27] In other words, talking about nature and transcendence is not primarily about finding metaphysical inspiration in a sunset or a walk in the park; the transcendence of nature is about agency that counters ecological destruction and creates the possibilities for the flourishing of life. An example from the world of undomesticated animals made the rounds during the raging wildfires in Australia in late 2019 and early 2020, where wombats shared their burrows with animals escaping certain death. As the myth had it, these wombats would actively shepherd other animals into the burrows to protect them, but such strong notions of agency are not necessary—the act of sharing itself is remarkable and appears to be a common way of life for wombats rather than some extraordinary action reserved for catastrophes.[28] Of course, even wildfires have some agency that can allow for the flourishing of new life. Such reflections on nonhuman agency might inspire conversations about ecological democracy in addition to economic democracy, but we cannot pursue them here.

Material practices are not limited to activities that are directly related to processes of production under the conditions of capitalism. They include, as Edwards suggests, "all those practices involving material bodies—organic and nonorganic—that . . . can be seen as a totality of practices that reproduce the relations of production over time."[29] This is an important reminder not only regarding nonhuman agency; human labor also needs to be seen in a broader light. At a time when labor is shifting and more and more people are pushed into the informal sectors, into casual jobs in the gig economy, temp jobs, or no formal jobs at all, production itself needs to be rethought, both in terms of its problems and its promises. This includes reproductive labor and any kind of work currently done without compensation, like housework, volunteer work, and the virtually uncompensated work of incarcerated people. What Edwards calls "the constitution of experience through the manifold forms of material practice outside the immediate space of production"[30] needs to be considered as well. This includes *"lo cotidiano,"* "the everyday," of which *Mujerista* theologian Ada María Isasi-Díaz used to remind theologians.[31] This brings a particular focus to the study of the immanent, and it transforms notions of what is considered transcendent, spiritual.

At stake is not merely analyzing the impact of immanent and material practices on religion; at stake is also the potential that reshaping processes of production and reproduction might have for reshaping religious experience and practice. This is not yet another romantic dream about life outside the dominant system—and it has even less to do with conventional understandings of transcendence. Even theologians might find themselves in cautious agreement with Edwards that "the material practices constitutive of modern life are the only grounds from which we could hope and expect to bring about important political and social transformations."[32] While material practices can and do make us compliant with the status quo—the world of productive and reproductive labor is designed to do exactly this under the conditions of capitalism—they also harbor the potential for resistance and for producing alternatives. This, then, is the place to start looking for transcendence again.

Which material practices are currently producing the most fertile ground for the alternative agency that is needed to transcend the exploitative relationships that affect both people and the earth? In the times of Jesus of Nazareth, the practices of peasants seem to have provided this ground; in Marx's time it was industrial labor—the so-called proletariat. Today, that question is more complex—some would point to the so-called precariat, which includes not only people belonging to the traditional working class but all whose existence is precarious now, even once-proud professionals and managers.[33] But the question is not just who is affected by capitalism but also who has what it takes to resist, and so movements of working people (both formal and informal) need to be taken into account, including the agency of nature and the environment that may well present one of the biggest challenges to capitalism yet.

Such an approach provides new clues for the study of religions like the Abrahamic ones, which originated among working people in times of economic and political turmoil and continued to be shaped in part by the struggle of working people throughout their history.[34] These and other religious movements developed in close relationships to working people and the material practices and the alternative ways of life and thought that grow out of them. This is also where progress in the fight against sexism (according to the traditions of socialist feminism) and racism (according to Martin Luther King Jr., W. E. B. DuBois, and many others) has been made, and here is where progress in the fight against ecological destruction finds increasing support, as well. At stake is not only the greening of the economy but the specific nature of the green economy, including green jobs that not only pay fair wages but also give workers a say in what is happening at work and where they, rather than the stockholders, are appropriating the profits for the sake of their communities. This is where economic democracy finds its roots, eventually impacting political, cultural, and religious democracy, as well.

In sum, as alternative religious subjectivities and practices emerge in the history of particular movements of exploited working people, in touch with the agency of the exploited earth, not only do notions of immanence and transcendence change; notions of religion and theology are changing, as well. Rather than trying to conjure up alternative religious subjectivities and practices out of thin air, scholars of religion and theology will have to study them while immersing themselves in the resistance movements of our time, exploring what all of this might mean for the future of religion, including its practices, its doctrines and beliefs, and its ways of life. To bring it down to a formula that sounds strange when pronounced at the level of the academic study of religion and theology: The confluence of materialism and religion needs movements, and it needs the movements of working people (formal and informal ones, self-organizing and cooperative), along with all the movements of nature and the earth. This is democracy at work at ever deeper levels—including economic and ecological ones—many of which we still must explore.

CONCLUSION

At the heart of my argument are questions of power and agency: what are we up against, where are the contradictions, and what are the alternatives? Among the most basic contradictions in neoliberal capitalism are still exploitative labor relationships, despite the developments of financial capitalism and other seemingly ethereal processes that have led scholars to argue that economics functions like religion. These exploitative relationships of production are compounded by the exploitation of the reproductive labor of women, minorities, and of nature itself. The tensions produced in these increasingly contentious relationships challenge the last pretenses of scholarly objectivity—the work of scholars of religion and theology cannot escape them, and so we need to decide what to do with them and how to rethink and reconstruct them.

To do that we will need to revisit fundamental questions like immanence and transcendence, religion and capitalism, and ecological catastrophes such as climate change that threaten the future of human life on the planet. Since power and agency do not fall from the sky (another widespread misunderstanding of transcendence), we will also need to investigate its formation and reformation in economic and ecological democracy and its embodiment in political, cultural, and religious forms of democracy.

In the meantime, there are already some explorations of how movements of working people, including women around the world and minority populations, shape and reshape ecological concerns, from community gardens to networks of worker cooperatives.[35] Is there anything else today that has the potential to transform the dominant status quo, both in its immanent and material functions on the ground and in its promises of transcendence? Challenging the logic of what might be called the Capitalocene rather than the Anthropocene, what can we truly say about the agency of people and the earth? In my view, this would be a most fruitful conversation, even though scholars of religion and theology may be a bit late to the game.

NOTES

1. This chapter is adapted from Joerg Rieger, *Theology in the Capitalocene: Ecology, Identity, Class, and Solidarity* (Fortress Press, 2022), 59–90.
2. Lynn White Jr., "The Historical Roots of Our Ecological Crisis," *Science* 155 (March 1967): 1205.
3. In addition to Clayton Crockett and Jeffrey Robbins, *Religion, Politics, and the Earth: The New Materialism* (New York: Palgrave Macmillan, 2012), see also the chapters in *Religious Experience and New Materialism: Movement Matters*, ed. Joerg Rieger and Edward Waggoner, Radical Theologies and Philosophies Series (New York: Palgrave Macmillan, 2015).
4. Sandra Harding, "Beyond Postcolonial Theory: Two Undertheorized Perspectives on Science and Technology," in *The Postcolonial Science and Technology Studies Reader*, ed. Sandra Harding (Durham, N.C.: Duke University Press, 2011), 21.
5. Diana Coole and Samantha Frost, "Introducing the New Materialisms," in *New Materialisms: Ontology, Agency, Politics*, ed. Diana Coole and Samantha Frost (Durham, N.C.: Duke University Press, 2010), 7.
6. Coole and Frost, "Introducing the New Materialisms," 29.
7. Coole and Frost, "Introducing the New Materialisms," 31.
8. See, for instance, Kathryn Tanner, *Christianity and the New Spirit of Capitalism* (New Haven: Yale University Press, 2019), and Harvey Cox, *The Market as God* (Cambridge, Mass.: Harvard University Press, 2016).
9. The term appears to have been coined by Jason W. Moore, *Anthropocene or Capitalocene: Nature, History, and the Crisis of Capitalism*, ed. Jason W. Moore (Oakland, Calif.: PM Press, 2016).
10. For the history, see Kevin M. Kruse, *One Nation under God: How Corporate America Invented Christian America* (New York: Basic Books, 2015).
11. Rosi Braidotti, "The Politics of 'Life Itself' and New Ways of Dying," in Cooke and Frost, *New Materialisms*, 214.
12. Crockett and Robbins, *Religion, Politics, and the Earth*, xx.
13. Jason W. Moore, "The Rise of Cheap Nature," in Moore, *Anthropocene or Capitalocene*, 91.
14. For slavery today, see Kevin Bales, *Blood and Earth: Modern Slavery, Ecocide, and the Secret to Saving the World* (New York: Spiegel and Grau, 2016).
15. Karl Marx, *Capital: A Critique of Political Economy*, vol. 1, book 1, trans. Samuel Moore and Edward Aveling, ed. Frederick Engels (Moscow: Progress, 1st English ed. of 1887), https://www.marxists.org/archive/marx/works/download/pdf/Capital-Volume-I.pdf.

16. Marx talks about these issues in *Capital*, 46–47, but since he does not mention the terms "exchange value" and "profit" in this section, it is difficult to follow.
17. Marx, *Capital*, 46–47.
18. Marx, *Capital*, 47.
19. Sara Ahmed, "Orientations Matter," in Coole and Frost, *New Materialisms*, 248–54.
20. Daniel Hartley, "Anthropocene, Capitalocene, and the Problem of Culture," in Moore, *Anthropocene or Capitalocene*, 162.
21. Jason Moore, "Rise of Cheap Nature," 93. Moore's argument that capitalism has been able to contain the rising costs of production by using nature's work as cheap resource, (114), is incomplete at best, as labor costs have also been kept low.
22. Jason Edwards, "The Materialism of New Materialism," in Coole and Frost, *New Materialisms*, 283.
23. To my knowledge, the notion of class as relationship rather than stratification has only been picked up very recently in religious studies; see *Religion, Theology, and Class: Fresh Conversations after Long Silence*, New Approaches to Religion and Power, ed. Joerg Rieger (New York: Palgrave Macmillan, 2013).
24. Karl Barth, *Dogmatics in Outline* (New York: Harper & Row, 1959), 40; for the background of Barth's life-long socialism, see the volume edited by George Hunsinger, *Karl Barth and Radical Politics* (Philadelphia: Westminster Press, 1976), especially the contributions by Friedrich Wilhelm Marquardt and Helmut Gollwitzer.
25. Joerg Rieger, *No Rising Tide: Theology, Economics, and the Future* (Minneapolis: Fortress Press, 2009).
26. Whitney A. Bauman, *Religion and Ecology: Developing a Planetary Ethic* (New York: Columbia University Press, 2014), 165.
27. See Bill McKibben, *Falter: Has the Human Game Begun to Play Itself Out?* (New York: Henry Holt, 2019), 12.
28. See "Viral Posts Claim Wombats Are Shepherding Animals to Their Burrows during Australian Bushfires," *My Modern Met*, https://mymodernmet.com/viral-wombat-post-australia-bushfires/.
29. Edwards, "Materialism of New Materialism," 283.
30. Edwards, "Materialism of New Materialism," 288.
31. See, for instance, Ada María Isasi-Díaz, *Mujerista Theology: A Theology for the Twenty-First Century* (Maryknoll, N.Y.: Orbis Books, 1996).
32. Edwards, "Materialism of New Materialism," 292.
33. Guy Standing, *The Precariat: The New Dangerous Class* (London: Bloomsbury, 2011).

34. See also Joerg Rieger and Rosemarie Henkel-Rieger, *Unified We Are a Force: How Faith and Labor Can Overcome America's Inequalities* (St. Louis: Chalice, 2016), and Ulrich Duchrow and Franz J. Hinkelammert, *Transcending Greedy Money: Interreligious Solidarity for Just Relations*, New Approaches to Religion and Power (New York: Palgrave Macmillan, 2012).
35. See, for instance, Kelsey Ryan-Simkins and Elaine Nogueira-Godsey, "Tangible Actions toward Solidarity: An Ecofeminist Analysis of Women's Participation in Food Justice," forthcoming; Jessica Gordon-Nembhard, *Collective Courage: A History of African American Cooperative Economic Thought and Practice* (University Park: Pennsylvania State University Press, 2014); and *Sustainable Lifestyles and the Quest for Plenitude: Case Studies of the New Economy*, ed. Juliet B. Schor and Craig J. Thompson (New Haven: Yale University Press, 2014).

Sacred Obligations: On the Theopolitics of Debt and Sovereignty

DEVIN SINGH

Debt and sovereignty are deeply intertwined. There are historical, theoretical, and practical correlates between the two. Historically, debt instruments appear on the scene in the ancient Near East at the same time as the emergence of forms of centralized sovereign governance. Theoretically, sovereignty can be thought of as exhibiting a debt-like structure, as it imposes forms of obligation on the bodies and populations that it manages and governs. Sovereignty and debt require one another conceptually. Practically, debt is a useful way to entrap and coerce and so serves as a tactic and strategy of sovereign control. Attending to such a nexus also invites attention to religion and theology. In the interstices between the center of authority and the bodies it seeks to govern, notions of the sacred are operative, inflecting theological discourse and religious practice, while legitimating or challenging the connections between debt and sovereignty. In this chapter, I pay particular attention to the purported tensions between debt and sovereignty—the ways the relation is posed as contentious even while being mutually sustaining. To this juncture I bring consideration of religious and theological dynamics.

We can observe these complex sets of relations in a variety of historical moments. Considering the debt dynamics of sovereignty, Diodorus of Sicily, a Greek historian writing in the first century BCE, looks back at the measures taken against severe indebtedness by the Egyptian kingdoms some 700 years earlier. As he writes:

> Citizens' bodies should belong to the state, so that the state can use the services *owed* it by its citizens in times of both war and peace. It would be absurd . . . for a soldier, perhaps at the very moment of leaving to fight for his country, to be taken away by a creditor for *debt*, and for private individual greed to thereby endanger the salvation of all.[1]

Diodorus is addressing the decision by the pharaoh Bakenranef (Gk: Bocchoris, ca. 725–20 BCE) to prohibit forms of debt that were guaranteed by the debtor's person—that is, debt that if not repaid could lead to debt slavery as collateral.[2]

Diodorus is concerned about contested allegiances. The citation contrasts obligations to the state or sovereign power—for he uses the language of "services *owed*"—with indebtedness to private creditors.[3] Indeed, he appears to set them into tension and competition, suggesting an agonism between these differing circuits of debt and obligation. The logic seems straightforward: if one is burdened and encumbered with debt to a fellow citizen, one is less readily available to render service to the sovereign, in particular as a conscripted soldier. More pointedly, if one is a debt slave to one's neighbor, one is removed from the tax circuit, having shifted from citizen-subject to captive property, thus eliminating one's status as taxable subject. The sovereign has clear interests in preventing or limiting the number of those falling into such subjection, since the community of potential soldiers and taxable subjects would be reduced.

Such tensions give a different glimpse into the much-idealized acts of Jubilee or debt cancellation attributed to the agrarian empires of the ancient Near East and as later codified in Hebrew scriptures. Kingdoms such as the Sumerians, Akkadians, and Babylonians bequeathed to civilizations in posterity the template and horizon—if not the actual practice—of debt relief measures proclaimed nationwide. Diodorus's reflection here, writing at a time when such kingdoms were already, for him, ancient history, shows us that certainly one potential motivator for Jubilee decrees involved a concern to maintain the relative free status of subjects for conscription—whether into military service or corvee labor for the nation—as well as for taxation. In other words, such Jubilee acts, while touted as merciful and benevolent proclamations of relief to all citizenry, may have had the purpose of insuring the availability of a subjected populace to do the bidding of the sovereign center.

I have argued elsewhere that such debt cancellation acts appear to function according to the logic of exception, as elaborated by theorists of sovereignty such as Carl Schmitt and Giorgio Agamben.[4] Ritually decreed at moments of accession, when a new emperor assumed the throne, such acts proclaimed the total supremacy of the ruler over all aspects of the governed territory, including its accounts and ledgers. In suspending the rule of debt, Jubilee actually proved it, releasing subjects from some debts to one another only to reinstate the debt of the subject to the sovereign.[5] Such a motivation can be glimpsed here in Diodorus as well, where at least one motive for clearing debts by the ruler was to ensure a freely conscriptable, governed populace.

What interests me in the present study is how, upon first glance, declarations of debt cancellation and the tensions that Diodorus highlights seem to pit the state and sovereign power against debt, severe indebtedness, and debt slavery. Such tensions fit a narrative of the state as savior, as order bringer, justice enactor, or equalizer, providing the liberative rule of law and curbing if not eliminating pervasive and corrosive forms of indebtedness in society. In such a tale, the sovereign provides relief from debt.

What I want to suggest is that the story is more complicated and that sovereignty and debt may not be as dichotomous or oppositional as they first appear. Certainly, we must account for the long-standing ways centralized state power is always triangulated amongst not only the governed but the aristocracy from which the ruler generally came. State interests and those of the ruling elite were typically aligned but may diverge at important moments. In the case of debt clearing, the sovereign act appears to threaten the right of wealthy creditors to claim their due.

Rather than being at odds, however, I claim that sovereignty operates according to the logic of a debt relation and that debt slavery, in particular, is constitutive of the logic of sovereign power as it has developed historically according to the state form. Sovereignty and debt slavery travel together. Sovereign states emerge as a particular kind of debt slavery, one that then seeks to eliminate competition from other would-be centers of credit with their own indebted clients or slaves. The sovereign is the most successful and expansive debt enslaver within a particular community and territory, one who effectively combines economic and political power with the logic of indebtedness and other forms of fealty-based obligation.

Such insights have implications for how we theorize the relation of the economic and the political today and what sorts of projects of state-supported debt relief we might champion or resist. Such archaeological retrieval also sheds light on why debt relations may be persistent and even increasingly widespread in society, accompanying as they do, I suggest, forms of sovereignty associated with notions of the state. Transcending debt relations and the indebted subject may be bound up with transcending sovereignty as the default structure of political relations today.

Such considerations have a theopolitical dimension, as well. While discussions of political theology have become firmly established such that many accept as commonplace the historical and conceptual intermeshing of theology and politics, conversations around economic theology are in some ways still nascent.[6] My aim in this chapter is both to help expose the ways debt and sovereignty travel together and reinforce each other and to bring considerations of religion and theological discourse to bear on this important link. Assessments of political theology will be improved when they consider the economic and economic theological dimensions typically at work.[7]

In this chapter, I recall the intermingling of debt practices and manifestations of sovereignty in early state forms. My claim is that sovereignty, understood here broadly as centralized authority and decision-making power associated with various kinds of states (empires, monarchies, republics), presumes and requires debt as well as forms of coerced labor that range from slavery to exploitation. I claim further that, when sovereignty, in the form of ancient Near Eastern kingship, becomes theologized in biblical tradition, with God decreed as king and the notion of law made divine, the mechanisms of debt are carried along on this cosmic journey. In scriptural imagination and resultant spiritualizations and theological formations, both the politics of kingship and the economic logic of debt are universalized and projected as part of the deep structure of the cosmos. While inquiries into political theology have long problematized (and capitalized upon) the genealogy of divine kingship, less attention has been devoted to the implications of economic ideas when transcribed into and developed from notions of the Godhead. This chapter is thus one step toward imagining and tracing the implications of this connection.

THEOPOLITICS AND THE PREPONDERANCE OF DEBT

This chapter is one discrete intervention within my larger project, which seeks to examine the pervasiveness and power of debt in contemporary society. I explore the senses of indebtedness that have been conveyed to the Western imaginary via religious and theological language and via forms of political institutionality that rely on and reproduce debt relations. Building on my previous work on money and the Western theological imaginary, I have been led to explore debt both because of money's deep interrelation with debt (for money is a token and record of debt) and because of the ways Jewish and Christian thought make extensive use of debt language and concepts.

The language of Jubilee, a creation of ancient Near Eastern empires that then appears in the Hebrew scriptures, becomes stylized and spiritualized in Second Temple Judaism and Christian traditions. Socioeconomic and political acts of debt forgiveness come to signify divine acts of moral forgiveness and redemption, adding a spiritual layer and enabling such ideas and ideals to be conveyed into new linguistic and institutional territories. Conversely, material-economic debt slavery comes to connote undesirable positions of spiritual or moral culpability before a divine sovereign and creditor and so get taken up in language about sin and guilt. In short, through such figurations, God becomes construed as a sovereign who issues credit and reclaims debt or who chooses to forgive, while humanity is presented as defaulting sin-debtors, unable to repay and in a state of bondage, longing for liberation.[8]

In previous work, I explored how the language of minting and coining, purchase, value, and monetary exchange reinforce and are in turn sacralized by Christian redemption language and theopolitical practice.[9] In this project, I am interested in how the nexus of economic and political acts around debt are given a type of divine authorization and sanction through their inclusion in biblical and ensuing theological tradition. Such inclusion also makes possible the conveyance of these ideas into new historical moments and contexts. We would not have movements like Jubilee 2000, Strike Debt, and other debt cancellation campaigns today if this language and these concepts had not been previously conveyed to the West via Jewish, Christian, and Islamic thought and tradition. Scripture, liturgy, moral exhortations, and theological doctrine act as repositories for these symbols and concepts that can be drawn upon in various eras to new and potent effect, even in ostensibly secular contexts.

This chapter is an effort to retrieve some of the sociopolitical context to language of debt slavery and debt cancellation in religious traditions. I am particularly interested in how such debt language works together with forms of sovereignty that have also been inherited in the long march of civilizational experimentation that starts in ancient Near Eastern and Indus Valley empires, was conveyed to Greece and Rome, and then was used as conceptual fodder for European identity construction and its mythologies of origins.[10] Might the pervasiveness of debt today and the speculations about debt-bound subjectivity in modernity, such as Maurizio Lazzarato's "indebted man," stem in part from the valorization and even divinization of debt relations in religious traditions and theological imaginations?[11]

Turning to premodern contexts and ancient sources provides a useful analytical and conceptual distance and paradigmatic disruption for the present. Such ancient contexts and ideas are useful to think with. I thus employ genealogical and archaeological techniques of excavating conceptual systems located within previous epistemes, systems that work part and parcel with various techniques and practices that arrange bodies and communities according to particular flows of power. A major concern of mine is bridging the gaps between the political, the economic, and the religious, gaps or disjunctions inserted in modernity, and instead reading these spheres together and always already in relation.

Perspectives from the ancient world are useful in this regard. Economic relations are less reified or "disembedded," in Karl Polanyi's terms, and ideological frameworks that legitimate the political realm draw explicitly on religious and theological ideas.[12] Political theology is thus centrally and clearly operative. What are the sacred and symbolic canopies that make particular institutional arrangements work? How might such conceptual systems have conveyed aspects of these ancient practices to new contexts and histories? How might have the ideological canopies not just fixed and enforced but shaped, acted back upon, and transformed new organizational and communal arrangements through their own complex evolution as bodies of thought, practice, and institution? These questions and others might be elucidated through a retrieval of the ancient symbolic imaginary bound up with some of the state form's earliest manifestations.

DYNAMICS OF DEBT SLAVERY

In addition to ancient practices of debt cancellation, thinking about the competitive or cooperative relation between sovereignty and debt raises debt slavery as a critical institution and trope to consider. Debt slavery is the shadow side to Jubilee, the condition from which Jubilee offers liberation. Debt slavery becomes a central metaphor for bondage to sin such that the language of redemption, so central to Christian discourse on salvation, is construed as a purchase or ransom out of spiritual or moral debt slavery. It is thus imperative to examine this institution and to differentiate it from other forms of slavery as well as think about its relation to sovereignty and other forms of power, authority, and communal formation.

In an article that helpfully documents the types and extent of debt slavery in various societies, anthropologist Alain Testart also considers the characteristics of societies that exhibit debt slavery practices.[13] He further explores the relation between the state and such institutions. Debt slavery is distinguished from chattel slavery and slavery through military conquest for several reasons. The first is the obvious financial basis for enslavement. Another distinguishing characteristic is that debt slavery is the one form of slavery permitted or at least practiced within a community, while the purchase of slaves or their conquest in battle are directed toward outsiders. This potential enslavement of one's neighbor through financial means is of course a major threat to communal solidarity. Yet, as Testart notes, such practices were widespread, both in state-based and stateless societies.

For a society to permit debt slavery, however contested it may be at times, such a society, claims Testart, must recognize relations of dependence as normal and acceptable. In other words, it must be constituted by hierarchies that include categories such as master and slave, owner and owned, and even patron and client.[14] Furthermore, it must be a society in which wealth plays an important role and indeed in which wealth can become correlated to and exchangeable with mastery, just as poverty becomes categorically exchangeable with slavery. These associations, which may appear natural to us, should be highlighted and made strange. For there is no inherent reason that wealth and mastery should coincide. Such a nexus reveals a society's particular relations, value systems, and constellations of power. Finally, Testart claims, a society in which debt slavery exists is a society that has facilitated a particular kind of power relation. As he writes, "Wealth in itself does not constitute power over persons; it only does so if certain institutions exist and are operative."¹⁵ Wealth, he notes, is firstly power over objects and resources and thus only indirectly and by secondary application power over people. Power to command people directly is political power, that of the sovereign, or the master over the slave. As Testart claims further:

> This explains why debt slavery rather than any other institution confers the greatest weight on economic power. It transforms what is essentially indirect power into direct power. . . . The institution of slavery already enabled the rich man to become a master, allowing him to command certain persons. Debt slavery goes further, enabling the rich man to bend the poorest and weakest persons of his society to his will (in addition to captive or purchased foreigners). This not only enlarges the sphere in which such a man may "recruit"; it extends his potential influence to the entire community, through the threat of possible slavery for reasons of debt or poverty that he can hang over the heads of some, and the protection he can give others against that same threat. It is in this way that he acquires clients.¹⁶

Debt slavery thus provides us with a crucial window into a particular nexus of economic and political power and gives clues as to how that fusion may permeate a constituted community. It helps show how the economic may fuse with and infuse the political, transforming the indirect power over others through objects and resources into the direct power over them as those very objects and resources. (This may also gesture toward the operations of biopower, where bodies and lives of the governed are those very resources that are the central objects of concern for sovereign power in regimes of governmentality.)

This melding of economic and political power explains the ambivalence of the relation between debt slavery and sovereignty. As Testart notes, the power of debt slavery "may be independent of all political control; it may be purely private. It may belong to a number of individuals. Society is then divided into a multitude of poles around which the poor, needy, guests, clients, and slaves gravitate."¹⁷ Such multipolarity within one community creates tensions and imbalances of power. In contexts of sovereign rule where centralization is key, intervention becomes necessary. He explains further:

The reasons for state intervention are obvious. A king's subject, a citizen of the polis, is neither subject or citizen if he becomes a slave. A slave has but one master. He pays no taxes, and owes no military service. Every time a freeman was taken into slavery, the political powers-that-be lost a source of fiscal revenue and a soldier. Debt slavery, in itself, together with the sale of one's person or kin into slavery, weakens the central power.[18]

Therefore, "the state has an obvious interest in limiting or abolishing debt slavery, and in most cases acts in accordance with that interest."[19] Hence Diodorus's insights considered previously.

So it would seem that state interests and the dynamics of debt slavery are not aligned. Debt slavery appears to compete for the loyalty and dependence of community members, potentially stealing them away from the sovereign. The state therefore has vested interests in curbing if not eliminating such practices. Does this then suggest a bifurcation and tension between sovereignty and debt slavery? Again, I think not. Testart likewise suggests not, and he points the way toward a rapprochement between sovereignty and debt slavery, although he does not go far enough.

Testart notes that there is a resemblance and even a possible genealogical relation between debt slavery and the practices of royalty. Indeed, we might view the type of direct power over clients amassed by creditors as a kind of proto-royalty. As Testart writes,

> In my view, the power of a man in control of considerable wealth, who already had numerous slaves in addition to his numerous descendants . . . clearly prefigures the power of the sovereign. He had the same direct power over his subjects as a king, and it hovered over the whole of the community, since everyone could ultimately be subject to it. Only analogous powers, in competition with his, could limit that power. But if a power struggle were instead to eliminate those other powers, how could that man not become the absolute master? In other words, debt slavery and sale of self or kin into slavery were factors that facilitated the emergence of royalty.[20]

Testart stops short of positing a connection here and seeks to keep these forms of power distinct. In part he does so through maintaining a strong dichotomy between political and economic power. He also leaves unaddressed the very real question of where one might get economic power without something like royal power: in other words, there is no apparent basis for one to claim possession of great wealth and have such claims respected by the community unless the claimant has some kind of political power. Claims to possession must be protected by communal tradition, sacred aura or charisma, or proto-law.[21]

Nevertheless, the blurring evinced in debt slavery is precisely the kind of blurring that appears at work in political communities organized by a sovereign center in which *political* fealty is bound up with *economic* concern for resources and value allocation and where one's access to goods and services is partly contingent upon one acknowledging the authority of that center. As we consider the appearance of states in the ancient Near East, what appears crucial is not merely the effective assertion of political power, force, or violence, but the skillful administration of resources and the employment of debt relations to manage a governed populace.

DEBT SLAVERY AND LABOR POWER

Debt slavery is not necessarily always aimed at extracting surplus value from the debtor in the form of interest and repayment. As Roland Boer claims, following Moses Finley, debt slavery in the ancient world functioned even more as a method to capture and command labor.[22] The primary resource concern of ancient regimes was human capital for labor. With high mortality rates and mobile populations, emerging empires sought ways to secure, corral, and contain a population for the sake of labor. Wars between sovereign centers of power rarely had as an aim the expansion into new territory or seizure of new land. Rather, the objective appears to be the capture and deportation of people, relocated around the Near East and Levant to toil on various work projects for the conquering state.

Boer also notes that exorbitant interest rates on loans appear calculated to make repayment nearly impossible in an effort to retain dependent labor. Such rates were still not fantastical, however, to maintain the ruse of repayability and retain the indebted laborer rather than force default and flight. Such predatory lending for the sake of entrapping borrowers to then be recruited as laborers appears to have been widespread and not just an extreme exception to more mundane lending practices. The destitute often had no choice but to enter into such agreements to receive a subsistence loan, resulting in peonage and enslavement for themselves or their family or both.

For Boer, then, we cannot exclude the centrality of labor in assessing debt slavery practices, and such arrangements had as one primary aim the recruitment and entrapment of a labor force via loan obligation. They had as their advantage the ability to draw forced labor from the ranks of one's own community rather than, as we saw, the attainment of chattel slaves or captured slaves from outside the community. Such mechanisms of forced labor were also able to be exploited by the well-to-do, those who perhaps could not afford the large up-front cost to purchase a chattel slave but who, through the more modest advance of goods as principal, could acquire a subjected laborer via the constraints of imposed debt obligation. This operation of debt slavery in the conscription of labor is critical because it was the securing and obligating of a labor force that was necessary to early state regimes as they arose in the ancient Near East.

INDEBTED LABOR AND THE RISE OF STATES

James C. Scott's *Against the Grain* examines the factors that needed to be present to form states, and his study assesses some of the earliest state formations in Mesopotamia.[23] He begins by noting that many of the factors classically associated with the rise of states, especially agriculture and sedentism, preceded states by tens of thousands of years. Early humans lived in communities and gathered in wetlands and other resource-rich areas, possibly even settling for a time until factors changed in the environment or they were attacked or driven out.[24] Fire was a chief domestication tool for millennia (discovered by *homo erectus* and bequeathed to *home sapiens*) that had a large impact on shaping human environments.

Scott's key claim is that states require appropriated and coerced labor, concentrated in discrete geographical spaces and sustained by grain. The attention to grain as imperial foodstuff is perhaps his most original claim and is the central theme of his book. Grain, Scott claims, exhibits a number of key characteristics that aid in hierarchical control and oversight, enabling states to rise and be sustained: grain (rice, wheat, barley, and millet) is countable, so that it can be measured, assessed, and most importantly taxed by imperial surveyors and accountants. It has distinct harvest cycles and is above ground and so can be monitored, timed, and then harvested at once, which allows it to be assessed as well as be pillaged by others. Scott notes that tuber empires would be very hard and frustrating to manage and tax, since tubers (such as yams and potatoes) are not visible above ground to be counted and can be harvested at varying times, allowing subjects to hide yields from surveyors. Grain, on the other hand, lends itself to observation, quantification, and regularized cycles of extraction. Grain was also durable after harvest, allowing for the accumulation of surpluses to be hoarded and managed over the long term. Centralized storehouses are thus observable in the archaeological record, serving as key sites of governance, with palace and temple elites monitoring the influx and disbursement of grain yields. As such, grain served as the central agricultural basis of command economies, one that contributed to particular political and economic formations.

The specific materiality of food in this case matters greatly for the types of sociality formed around it. Grain as enumerable, quantifiable, durable, and able to be tracked and cataloged creates a scope of possibilities for human response. More than this, we might consider the forms of agency exerted here by the land and its flora as more than a passive recipient of human action and rather as offering a range of possible responses by human actors in this case.[25] As Scott observes, we have no evidence of tuber empires. The materiality of grain matters. The land is a co-agent in an extensive and diffuse network of human and nonhuman actors, contributing to the shape of human sociality and the course that civilization takes.

We also see a parallelism among the enumeration, quantifiability, and domestication of grain with similar tactics carried out on domesticated animals and human laborers. All three spheres of life (agriculture, animal husbandry, and human labor) are disciplined according to such principles within a system of control, coercion, extraction, and concentration of resources seen in this period of history. Whether or not we agree with David Graeber that human slavery was a key origin of quantification, as the sort of zero point or absolute standard that enables comparisons among other regimes of value, there is a relation between the capture and enslavement of humans in these early empires and the quantification and sequestering of plants and animals.[26] There are a relationality, resonance, and correspondence here that appear significant in getting this shape of life off the ground. It is a set of parallelisms that of course continue in various forms through to our present.

Scott notes that the move from hunting, gathering, and pastoralism to one of stationary agriculture is not obvious. It is not self-explanatory that one would want to settle down to what he sees as the drudgery of sedentary and immobile cultivation. There are also disease and epidemic concerns as a result of close-quarter living that do not affect mobile groups. Plus, there is the fragility of dependence on single food sources as well as potential for land not to sustain the crops, for disease, blight, or climate factors to eliminate an entire harvest. For these reasons, he claims, the move to grain-based regimes required the capture, securing, and coercion of labor.

Scott recalls that slavery was a key institution in the ancient world needed to ground and support these centralized regimes. As Scott notes, "Bondage appears to have been a condition of the ancient state's survival."[27] Slaves were caught in battle, invasions, or conquests of neighbors or bought by mobile "barbarians" who sold their own or captured and traded others. We need to consider how debt was one tactic that eventually became central, perhaps because of the justifications it provides, the window dressings of legitimacy and apparently socially accepted and enforced conventions. What is clear is that fugitivity was everywhere and prevalent. The ancient poor peoples in the region were constantly trying to get away from these slaving and coercive centers and the independent traders and marauders who sourced them.

The purpose of concentrated settlements such as the plantation and early state centers was to "produce a surplus available to nonproducing elites."[28] As Scott writes,

> A peasantry—assuming that it has enough to meet its basic needs—will not automatically produce a surplus that elites might appropriate, but must be compelled to produce it. Under the demographic conditions of early state formation, when the means of traditional production were still plentiful and not monopolized, only through one form or another of unfree, coerced labor—corvée labor, forced delivery of grain or other products, debt bondage, serfdom, communal bondage and tribute, and various forms of slavery—was a surplus brought into being.[29]

Debt slavery thus finds an important place alongside various forms of slavery and coerced labor used to generate surplus resources to be amassed, catalogued, enjoyed, and reallocated by the governing center.

In fairness, Scott at times glorifies free barbarian life, in contrast to life in ancient cities, as if the impulse to gather together and work out a means of group governance is always externally imposed. If humans concentrated together and chose sedentary agricultural life instead of the glorious proto-biker "life on the road" that Scott depicts of roaming tribes, then someone must have forced them.[30] Yet, it is highly probable that early humans saw the benefits of large group life together, of agriculture, as well as of the ameliorative potential for lending to one another in times of need and actually stipulating a return at a later point (i.e., not as vague gifts but explicit loans). Through the complex push-pull of internal negotiations and administration, so-called "upstarts" could gain control and become authority figures or otherwise gain greater shares of power.[31] In other words, there are arguably benefits to large population concentrations, centralized forms of rule, and types of lending that get missed in the typically negative portrayals (including my own) of both debt and sovereignty.

SACRIFICE AND DEBT IN THE TEMPLE-PALACE COMPLEX

Beyond debt slavery as a discrete and key institution for consideration, indebtedness writ large characterizes conditions under sovereignty. While much could be said about this here, what I want to recall in short is that submission under centralized power (and to be sure under other types of authority) is often construed as a debt, as services owed, even while being marked by actual economic links, where taxes are paid or where tithes and tributes are offered. These are distinct conduits of obligation but are regularly blurred. In other words, following Testart, being politically bound to a master or king is a different sort of obligation than being bound to a lender. Yet sovereign forms of rule often combined the political with the economic. The previous excerpt from Diodorus shows this elision as he moves from services owed to the sovereign (such as military service) as somehow in competition with monetary debts owed to local creditors. Again, these are distinct logics, but the language of debt to political power allows their mutual implication, and it is a link that is underscored by actual economic obligations to the governing center.

As economic historian Michael Hudson has extensively shown, it was the production, administration, and allocation of grain surpluses that gave rise to tabulation and quantification practices by the temple-palace complex in these Mesopotamian regimes. Debt relations were marshaled on a variety of levels to motivate such a system. In addition to the coerced indebted labor used to plow and harvest fields owned by the temple-palace, the rendering of such yield back to the governing center was also construed as an obligation, as a debt owed to the king for his sacred oversight of the populace, and to the priests for their administration to the gods and mediation of favor to the community.

Faced with the influx of grain, the temple administrators needed to develop ways of measuring, tracking, and reallocating such grain among the governing elites and as rations to the laborers providing the grain, the civil engineers devising new irrigation tactics for the fields, and the soldiers providing protection for the state and its land and resources. Here for the first time in the historical record we find the enforcement of standardized measurements.[32] As Scott also explains, "The intention was to eliminate a host of local, vernacular, and idiosyncratic practices of measurement so that, for the first time, the ruler at the center could have a clear view of the wealth, production, and manpower resources at his disposal. It aimed at creating a centralized state rather than merely a strong city-state that was content to extract occasional tribute from a constellation of quasi-independent satellite towns."[33]

Such tabulation practices, essentially appearing first as scratches on clay tablets, also led to the written cuneiform practices in these empires in Mesopotamia in the 3000s BCE. It took centuries before these forms of proto-writing were used to represent speech and even longer before they were used for capturing ritual language, poetry, and other aspects of literary culture. Writing was first and foremost a quantitative technique of accountancy. As Scott notes, "The first writing was . . . an artifact of state building, concentration of population, and scale. It was inapplicable in other settings." Some have even suggested that "writing was elsewhere resisted because of its indelible association with the state and taxes, just as ploughing was long resisted because of its indelible association with [forced] drudgery."[34]

With such written tabulations and with the quantifying practices of enforced resource extraction, we have the tools in place for the development of both legal codes and money, both central arms of administration and governance in states. Indeed, much of Hudson's work explores the conceptual prehistory of money as first emerging as a tactic for imposing and administering debt among these ancient populations. Long before it takes the form of coins in Greece millennia later, money appears on the scene in these ancient agrarian regimes. It functioned as a conceptual tool for creating proportions between disparate goods, relating them to an official standard of measure imposed by the sovereign, and further demarcating all goods in terms of a percentage of tax or tribute owed to that sovereign center.

In regard to money's appearance in Greece as coins, one prominent theory that relates here has been advanced by Bernhard Laum and taken up by W. H. Desmonde, Richard Seaford, and Alla Semenova.[35] Looking at archaic Greek practices of communal sacrifice, such theorists posit that coinage first emerges as a metallic token signifying one's participation in the ritual. Where some early Greek sacrifice involved bringing cattle to be slaughtered by the priest and distributed among the participants, metal tokens symbolizing the iron spits upon which the flesh was roasted gradually came to stand in for distributed meat. Participants could receive a symbol that designated their portion and that in turn provided evidence that they had donated or participated in this central ritual of state. The point in such theory is to emphasize that coinage may have emerged out of a complex system of religiopolitical debts to gods and their representatives (priests/chieftains), rather than in "profane" market exchanges as a way to make barter less inconvenient (as the "Just-So" story of classical economists goes).

Here ritual and religious obligation overlap with subservience to political powers. Debts to the gods and kings are blurred, as well as marked, quantified, and tracked. Rather than a blurring, it is perhaps more accurate to say that at such a stage in history such realms and practices were yet undifferentiated, such that obligations to gods and kings were of a piece. While such observations focus on archaic Greece in the early first millennium, they may also provide clues to patterns that emerged millennia earlier in the ancient Near East. Given the existence of a temple-palace complex that received grain offerings, it is conceivable that such offerings were motivated not simply out of political fealty but conceived of as tithes to divine figures. The resultant need to track, count, tabulate, account for, and reallocate such offerings then provides context for accounting and monetary breakthroughs. The long-standing elision of debt and sacrifice in the religious imagination may thus be present from these early moments.

THEOLOGIZATION OF SOVEREIGNTY, LAW, AND DEBT

Forms of debt bondage aid sovereign power, and patterns of debt management and distribution appear constitutive of the governance of bodies that sovereignty carries out. Sovereignty thrives on the skillful administration of debt: both the allocation and redistribution of the debts of the governed. In so doing it also masks and deflects its own debts, reframing them as debts of the populace, debts borne by the collective for which the sovereign center functions as representative. Sovereign power as a center of authority and final decision, overseeing and managing a population through various arms of governance and extracting value from such a population for its own sustenance, operates within debt relations. Debt slavery enables the conscription of labor and the concentration of wealth in the form of goods and human living labor power. It uses the mores and traditions of communities to enforce these relations, not through capture and open violence as in chattel slavery, but through the language of economic convention, of credit extended and debts owed. Neighbors and community members fall into the orbit of centers of economic power, which can then command political power over their clients.

More successful centers of political and economic power work through various enslavement practices to coerce labor and produce resource surpluses, which the governed then render to the center based on additional notions of moral indebtedness. These debt relations become encoded in law and normalized through ritual and tradition. Other persistent and competitive forms of debt slavery within the governed population are now seen as threats to the sovereign center. They are thus called out and made strange, further masking the centrality of debt operations in the very sovereign center that is now issuing debt cancellation edicts and inveighing against debt slavery. Sovereignty's own reliance upon debt and its relation to archaic debt slavery are forgotten, and the state becomes the protector of the weak and vulnerable, who are exploited by ruthless and predatory private lenders. In this way the political can further distance itself from the economic and hide its own economy, rendering natural its attempts to monopolize the debt relation.

This image of the sovereign, the ancient king or emperor, as defender of the weak and powerless was incorporated into religious language to describe the gods. Eventually it also shaped early biblical tradition as the gods receded into a focus on one supreme God, Yahweh. The implications of the incorporation of such imagery are vast, and much has been written about this watershed transition. Of the many themes that arise in such discussions, I want to recall the purported significance of decreeing God as king and making law divine. Ascribing royal imagery to God has a paradoxical impact. On one hand, it legitimates kingship to a profound degree by making kingship a property of God and thus making kingship a metaphysical claim, one that permeates the cosmos. God as creator, whose creation reflects the imprint of the divine, is also a king, thereby making earthly kingship a legitimate echo and replication of godly rule. On the other hand, it relativizes and has the potential to undermine all earthly kingship by declaring everyone as subject to the one true, supreme king. Kingship may be reserved for God alone. Both traditions and implications are seen in ancient Israel.[36]

Furthermore, by making law divine, as a dictate directly from the mind and heart of God, such traditions can also elevate law to a divine, cosmic status while relativizing earthly or political law.[37] In other words, the paradoxical pattern emerges here as well. Kings may claim their laws have authority in light of divine law, or kingly attempts at legislation may be sidelined in favor of such eternal decrees. Furthermore, legal traditions in the ancient Near East were typically tied to the will and personality of the king, seen as reflections of their own justice. As such, law was transitory and tied to individual moments of sovereignty and exception. Making law divine allows a transcendentalizing of the code, making it stable over transitions in leadership and allowing its attributes to be transferred among periods and territories in a way less possible when tied to local regents.

While there is much to say about these two impulses (and, indeed, books have been written on them), I invoke them to bring to mind the ways that debt travels alongside here. Debt appears implicated in kingly assertions of authority through the mechanisms of resource centralization and control. Debt also appears significant in declarations of kingly benevolence through Jubilee. Debt is present in law codes regulating who may transact with whom according to its logic and who may or may not be exploited and upon what terms. What happens when such links enter heaven? What happens when God becomes the one who cancels debt but who also holds God's people accountable to it? What happens when earthly laws about debt relations become moral, spiritual, and cosmic laws about guilt and recompense? In short, what happens when God is not only king but also creditor and sovereign debt lord?

Ancient links between debt and sacrifice also set the stage to construe sacrifices to God as repayment of debt or as otherwise implicated in some type of economy. Sacrifice need not be understood economically, and indeed many forms of sacrifice prioritize relationship and communal experience, for instance. But the blurring with debt allows the notion of sacrifice as compensatory in some way, introducing economy into the equation. Thus, the mingling of these practices in ancient regimes paves the way for sacrifice, guilt, and debt to also be elided in biblical tradition, with a diverse set of consequences.[38]

I conclude with Robert Bellah's claim about the influence of the Hebrew scriptures upon subsequent religiopolitical formations in light of Bellah's preceding discussion of kingship and divinity:

> Taken as a whole, the historical framework of the Hebrew Bible is metanarrative big time, to be sure, but a metanarrative that no culture that has received it has ever been able to escape. And a metanarrative powerful and flexible enough so that movements and countermovements, establishments and heresies, could all turn to it to justify ethical/social/political programs, programs that would contribute, not always to the good, to all subsequent historical dynamics.[39]

One may want to quibble with aspects of this, but I accept the main thrust about the overwhelming influence of Hebraic notions of divine kingship and polity on subsequent political developments, particularly in the "West." Contained but unspecified in studies such as Bellah's are the ways debt (and of course other economic) practices also contribute to a certain set of historical legacies. The economic must be thought together with such theopolitical trajectories, given how essential the economy is to politics and how imbricated economic ideas, metaphors, and practices are with religion. My moves in this chapter have attempted to insert these economic considerations into the conversation, to recall that, inasmuch as divine kingship has left its imprint on medieval and even modern conceptions of sovereignty and governance, the divine economy and its heavenly creditor have also left traces that contribute to debt's ongoing centrality and influence in Western societies today.

NOTES

1. Diodorus of Sicily, *Bibliotecha Historica* 1.79.3, emphasis added; referenced in M. I. Finley, *Economy and Society in Ancient Greece* (New York: Viking Press, 1982).
2. Diodorus's account is contested. On the Egyptian records that do not corroborate his, see Tomasz Markiewicz, "Security for Debt in the Demotic Papyri," *Journal of Juristic Papyrology* 35 (2005): 141–67; Tomasz Markiewicz, "Bocchoris the Lawgiver—or Was He Really?," *Journal of Egyptian History* 1, no. 2 (2008): 309–30.
3. While we might be tempted to describe the obligation to the sovereign here as "debt," I would argue that "fealty" or some other term denoting political obligation is preferable to avoid collapsing political and economic obligation. To the extent that the subject must render actual economic value back to the sovereign, such as taxes and tribute, this may be depicted as debt. Of course, Diodorus—as most of us tend to in conventional speech—collapses these distinctions here.
4. Carl Schmitt, *Political Theology: Four Chapters on the Concept of Sovereignty*, trans. George Schwab (Chicago: University of Chicago Press, 2005); Giorgio Agamben, *State of Exception*, trans. Kevin Attell (Chicago: University of Chicago Press, 2005).
5. Devin Singh, "Debt Cancellation as Sovereign Crisis Management," *Cosmologics*, January 18, 2016, http://cosmologicsmagazine.com/devin-singh-debt-cancellation-as-sovereign-crisis-management/; Devin Singh, "Sovereign Debt," *Journal of Religious Ethics* 46, no. 2 (2018): 239–66; Devin Singh, "Exceptional Economy: Sovereign Exchanges in Carl Schmitt and Giorgio Agamben," *Telos* 191 (Summer 2020): 115–36.
6. To remedy this gap, see now Stefan Schwarzkopf, ed., *The Routledge Handbook of Economic Theology* (London: Routledge, 2020).
7. I attempt to demonstrate the importance of examining this nexus of theology, politics, and economy in Devin Singh, *Divine Currency: The Theological Power of Money in the West* (Stanford, Calif.: Stanford University Press, 2018).
8. On actual ancient practices of debt and debt slavery, see, e.g., Gregory Chirichigno, *Debt-Slavery in Israel and the Ancient Near East* (Sheffield: JSOT Press, 1993). On the metaphorization of debt in the Hebrew Bible, see Gary A. Anderson, *Sin: A History* (New Haven: Yale University Press, 2009).
9. Singh, *Divine Currency*.
10. In other words, I resist a classicist impulse that asserts Europe as the clear descendent of these regimes and claim rather that European thinkers intentionally marshaled a sense of classical heritage in their own self-construction of European identity; see, e.g., Dipesh Chakrabarty, *Provincializing Europe:*

Postcolonial Thought and Historical Difference (Princeton: Princeton University Press, 2000); Moira Fradinger, *Binding Violence: Literary Visions of Political Origins* (Stanford, Calif.: Stanford University Press, 2010); Cedric Robinson, *An Anthropology of Marxism* (London and New York: Ashgate, 2001).

11. Maurizio Lazzarato, *The Making of the Indebted Man: An Essay on the Neoliberal Condition*, trans. Joshua David Jordan (Los Angeles: Semiotext[e], 2012).
12. Karl Polanyi, *The Great Transformation* (New York: Rinehart, 1944); Karl Polanyi, ed., *Trade and Market in the Early Empires* (Glencoe, Ill.: Free Press, 1957).
13. Alain Testart, "The Extent and Significance of Debt Slavery," *Revue française de sociologie* 43 (2002): 173–204.
14. See also Kent V. Flannery and Joyce Marcus, *The Creation of Inequality: How Our Prehistoric Ancestors Set the Stage for Monarchy, Slavery, and Empire* (Cambridge, Mass.: Harvard University Press, 2012).
15. Testart, "Debt Slavery," 195.
16. Testart, "Debt Slavery," 196.
17. Testart, "Debt Slavery," 196.
18. Testart, "Debt Slavery," 197.
19. Testart, "Debt Slavery," 199.
20. Testart, "Debt Slavery," 200.
21. On the possibilities of pre-political sacral kingship, see Marshall Sahlins, "Kings before Kingship: The Politics of the Enchanted Universe," in *Sacred Kingship in World History: Between Immanence and Transcendence*, ed. A. Azfar Moin and Alan Strathern (New York: Columbia University Press, 2022), 31–52.
22. Roland Boer, "Biting the Poor: On the Differences between Credit and Debt in Ancient Israel and Southwest Asia," *Journal of Religion and Society* 16 (2014): 1–21. See also Finley, *Economy and Society in Ancient Greece*.
23. James C. Scott, *Against the Grain: A Deep History of the Earliest States* (New Haven: Yale University Press, 2017).
24. Indeed, a refrain in Graeber and Wengrow's recent intervention is that for millennia humans explored temporary forms of governance, shifting from hierarchical to consensus-based rule and back as circumstances and needs shifted. The narrative of linear development from simple egalitarian to complex hierarchical empires is thus facile; see David Graeber and David Wengrow, *The Dawn of Everything: A New History of Humanity* (New York: Farrar, Straus and Giroux, 2021).
25. I have in mind here—though in a completely different context—explorations of the land's agency and participation in the violence of U.S. border policing, as movingly explored in Jason De León, *The Land of Open Graves: Living and Dying on the Migrant Trail* (Oakland: University of California Press, 2015).

26. See David Graeber, *Debt: The First 5,000 Years* (Brooklyn: Melville House, 2010). For critiques of Graeber's origin story of slavery, quantification, and debt, see Fiona Allen, "Money, Debt, and the Business of 'Free Stuff,'" *Rethinking Money, Debt, and Finance after the Crises*, ed. Melinda Cooper and Martijn Konings, *South Atlantic Quarterly* 114, no. 2 (2015); Bill Maurer, "David Graeber's Wunderkammer, *Debt: The First 5000 Years*," *Anthropological Forum* 23, no. 1 (2013): 79–93.

27. Scott, *Against the Grain*, 30.

28. Scott, *Against the Grain*, 151.

29. Scott, *Against the Grain*, 152.

30. This connects with Marshall Sahlins's observation that origin stories of many kings and "big men" present them as external conquerors. Even archaic human communities had trouble accepting that they, themselves, may have chosen a king, opting instead to see them as invaders or usurpers or as externally imposed. See Marshall Sahlins, *How "Natives" Think: About Captain Cook, for Example* (Chicago: University of Chicago Press, 1995). Arguably this ambivalence is reflected in the Hebrew scriptures as well, with the people imploring a resistant Samuel to appoint them a king, followed by Israel's (and Yahweh's—as proxy for Israel) profound dissatisfaction with Saul.

31. This process of relatively egalitarian governance and decision making often giving way to centralized and dynastic authority is richly documented in Flannery and Marcus, *Creation of Inequality*.

32. Michael Hudson, "The Archaeology of Money: Debt versus Barter Theories of Money's Origins," in *Credit and State Theories of Money: The Contributions of A. Mitchell Innes*, ed. L. Randall Wray (Northhampton, Mass.: Edward Elgar, 2004); Michael Hudson and Marc Van de Mieroop, eds., *Debt and Economic Renewal in the Ancient Near East* (Bethesda, Md.: CDL Press, 2002). On a comparable case in ancient Egypt, see John F. Henry, "The Social Origins of Money: The Case of Egypt," in *Credit and State Theories of Money: The Contributions of A. Mitchell Innes*, ed. L. Randall Wray (Cheltenham, UK and Northampton, Mass.: Edward Elgar, 2004).

33. Scott, *Against the Grain*, 145.

34. Scott, *Against the Grain*, 148.

35. Bernhard Laum, *Heiliges Geld: Eine historische Untersuchung über den sakralen Ursprung des Geldes* (Tübingen: Mohr, 1924); W. H. Desmonde, *Magic, Myth, and Money: The Origin of Money in Religious Ritual* (New York: Free Press of Glencoe, 1962); Richard Seaford, *Money and the Early Greek Mind: Homer, Philosophy, Tragedy* (Cambridge: Cambridge University Press, 2004); Alla Semenova, "Would You Barter with God? Why Holy Debts and Not Profane Markets Created Money," *American Journal of Economics and Sociology* 70, no. 2 (2011): 376–400.

36. On these themes, see Robert Bellah and Hans Joas, *The Axial Age and Its Consequences* (Cambridge, Mass.: Belknap Press of Harvard University Press, 2012), 59–76; Robert N. Bellah, *Religion in Human Evolution: From the Paleolithic to the Axial Age* (Cambridge, Mass.: Belknap Press of Harvard University Press, 2011); Marc Zvi Brettler, *God Is King: Understanding an Israelite Metaphor*, JSOT Supplement Series (Sheffield, UK: Sheffield Academic Press, 1989); Eckart Otto, "Political Theology in Judah and Assyria: The Beginning of the Hebrew Bible as Literature," *Svensk exegetisk årsbok* 65 (2000); Stephen A. Geller, "The God of the Covenant," in *One God or Many? Concepts of Divinity in the Ancient World*, ed. Barbara Nevling Porter (Chebeague, Maine: Casco Bay Assyriological Institute, 2000); Stephen A. Geller, *Sacred Enigmas: Literary Religion in the Hebrew Bible* (London: Routledge, 1996).

37. On these themes, see Michael Walzer, *Exodus and Revolution* (New York: Basic Books, 1985); Regina Mara Schwartz, "Law and the Gift of Justice," in *Crediting God: Sovereignty and Religion in the Age of Global Capitalism*, ed. Miguel E. Vatter (New York: Fordham University Press, 2011); Jan Assmann, *Moses the Egyptian: The Memory of Egypt in Western Monotheism* (Cambridge, Mass.: Harvard University Press, 1997); Jan Assmann, *The Price of Monotheism* (Stanford, Calif.: Stanford University Press, 2010); Jan Assmann, *Religion and Cultural Memory: Ten Studies*, trans. Rodney Livingstone, Cultural Memory in the Present (Stanford, Calif.: Stanford University Press, 2006). For a summary of some of the issues in Assmann's program, see Daniel Steinmetz-Jenkins, "Jan Assmann and the Theologization of the Political," *Political Theology* 12, no. 4 (2011): 511–30. I focus in this chapter on debt cancellation in ancient Near Eastern empires preceding Israel, concepts and practices that the Hebrew scriptures inherit. It merits debate about whether Jubilee acts could be practiced in Israel's nonmonarchical period and, if so, whether they would follow a similar sovereign logic of exception around a centralized authority or governing group or promote something much more "democratic."

38. This elision is explored in Anderson, *Sin*. Yet he focuses narrowly on the influence of Aramaic on the Hebrew scriptures during the Persian exile and does not attend to this much longer history and broader sociopolitical context that precedes the exile and characterizes most of the ancient Near Eastern societies where the biblical texts emerged. For reception of such themes among early Jewish Christ followers, see Nathan Eubank, *Wages of Cross-Bearing and Debt of Sin: The Economy of Heaven in Matthew's Gospel* (Berlin: De Gruyter, 2013).

39. Bellah, *Religion in Human Evolution*, 322–23.

❧ Curating Futures: The Curatorial as a Theological Concept

DANIEL A. SIEDELL

The task of curating is to make junctions, to allow different elements to touch.
—HANS ULRICH OBRIST

Curators have always been a curious mixture of bureaucrat and priest.
—DAVID LEVI STRAUSS

To think the future theologically in relationship to what Catherine Keller refers to as the "triple creeps of the apocalypse" stresses existing categories of thought, even those built with the "transdisciplinary" scaffolding that many of us utilize to do our work.[1] But this holy trinity of catastrophes—ecological, political, and economic—as this volume implies by its title, might require even more flexible conceptual apparatuses. Our futures—both possible and impossible—just might depend on our capacity to imagine ecology, politics, and economics in new, more flexible ways. The challenges that face us by the economic-ecological-political apocalypse is that they are not merely three manifestations of a single catastrophe. Moreover, intersectional thinking presupposes connecting hubs. But often these economic, ecological, and political catastrophes *don't meet*—their gravitational forces pulling them apart as often as they connect or intersect. How can we think in the light of this impossible, contradictory, and irreconcilable multiplicity, always moving, unfolding, resistant to being *grasped*? Philosopher of science Isabelle Stengers describes it in this way when she writes, "This new situation doesn't signify that the other questions (pollution, inequalities, etc.) move to the background. Instead, they find themselves correlated, in a double mode."[2] How are we to think this correlating situation that accommodates its ever intensifying, increasing, unceasingly unfolding multiplicities and intensities related to the many crises, catastrophes, and apocalypses through which we are living? Where can we find the creative resources to respond theologically with the imagination necessary either to meet this catastrophic moment with faith, hope, and love or to reveal their vibrations? How can we "assemble" such ever increasing, expanding, "creeping" catastrophes, not from the privileged wilderness margins of prophetic distance but *in media res* and amongst the agents, practices, and concepts that are compromised, complicit, and irrelevant? Eco-theorist and activist Timothy Morton argues that one place to look for such resources is the visual arts.

In *The Ecological Thought*, Morton writes, "Studying art provides a platform, because the environment is partly a matter of perception. Art forms have something to tell us about the environment, because they can make us question reality."[3] Moreover, art is "a place in our culture that deals with intensity, shame, abjection, and loss. It also deals with reality and unreality, being and seeming."[4] The work of art, Morton argues, offers the potential to reveal and forge relationships, explore the vulnerabilities of emotion and intimacies of experience, and encourage imagining new possibilities. Works of art need not have "the environment" or "ecology" as their subject to *be* ecological and serve as an important companion for ecological thinking. Indeed, Morton writes, "Art is ecological insofar as it is made from materials and exists in the world."[5] Ultimately, for Morton, *all art is ecological* because it "disables getting from A to B by causing the illusion of smooth functioning to malfunction, so as to reveal the spooky openness of things."[6] For Morton, ecological activism requires that we recognize and experience this "spooky openness."

Morton's argument *feels* right to me—what Morton calls *truthfeel*, especially since I'm trained as an art historian and art museum curator and have spent my adult life in front of works of art in museums, galleries, and artists' studios. This dimension of art on which he draws important attention needs to be explored in more depth and in new ways not only by eco-theorists, but also by constructive theologians. However, there can be a tendency to idealize and mythologize artistic practice and creativity that can serve to entrench ways of thinking and practice that can rhyme with the myths of a transcendent and sovereign ex nihilo monarchical creativity that is always already implicated in and in some ways responsible for the very political, ecological, and economic crises that fuel neoliberal innovation.[7] And for Morton, this is a distortion of art.

> Art only half works as a human-scaled bourgeois ideology reproduction device if you put just a tiny drop of it in-to the soup, and don't examine it too carefully or treat it as decoration.[8]

The impact of the arts and its potential for transformational thinking and practice, to which Morton draws attention, I argue, is thus not located in the artist's studio, in one's creativity, but in *our* experience of and relationships we forge with the objects, artifacts, events they generate. It is the work of art's affective, experiential, and cognitive effects—the "lines of flight" and "intensities" of feeling it catalyzes in art's reception, its effect on viewers (listeners, spectators, readers).

However, we experience works of art—develop relationships with them—primarily (whether directly in person or virtually) in and through *curatorial practices*—that is, through curated exhibitions in public spaces, like art museums and galleries. This public and social platform relies on, introduces, and releases a legion of political, ecological, and economic energies—including some even at odds with our goals.

It is the figure of the curator, rather than the artist, then, that I would like to explore in this chapter. The image of the curator and the concept of the curatorial exemplify a creative practice that is less visible, more modest, and perhaps a bit more easily disentangled from the romantic myths of the Artist as sovereign Creator ex nihilo while at the same time is more transparently implicated in the very political, economic, and racial structures that generated this crisis. This moment requires *curation* rather than *creation*. In fact, I would like to suggest that to think anything theological at this moment requires something like a curatorial practice. It requires a creative assembling that does not dismiss but invites and works with those actors, ideas, and practices that are complicit, compromised—that are brought into new relationships, new juxtapositions, as we curate theological exhibitions of our possible and impossible futures.[9]

I discern the curatorial in a short yet provocative essay by Morton entitled, "Queer Green Sex Toys." He writes, "Ecological reality is contradictory, and if we humans want to go about 'saving,' 'preserving' or as I'd prefer to say, curating it, we had better allow some things to contradict themselves or else."[10] What does it mean to curate this emerging ecological reality? For Morton, to curate provides a way to work with the multiple—and contradictory—dimensions of this reality—Stengers's "new situation"—that resists the religious metaphors and images of too little agency (piety or fideism—God will take care of it) or the scientific figures that suggest too much agency (science or politics will solve the "problem"). Both religious and scientific images can encourage an approach to these catastrophes from a distance—from a perspective—that reifies. Yet the problem, as Keller points out, is that they don't stand still for us to capture in thought. They creep (when they don't lurch or scatter off into different directions.) And moreover, they cannot be observed on the horizon. They are here, now, with us. As Morton, the end of the world is already here. We are living with these catastrophes and implicated as they continue to unfold, mutate, under our feet, shifting our foundations and destabilizing our thought and practice. The image of "curator" and the concept of "curating" offer helpful ways for theology to incorporate this incessant movement and multiplicity into its practice, a practice of mediation—of assembling, construction.

The vibrations of the curatorial are felt in the recent work of Catherine Keller. In *Facing Apocalypse: Climate, Democracy, and Other Last Changes*, Keller suggests that the powerful metaphor of apocalypse (she prefers to call it a "metaforce") must be faced to "crack open, to disclose" the "forces of destruction" that have underwritten "our economic habits, democratic disarray, and ecological suicide."[11] Like Morton, Keller reaches for the metaphor of the curatorial. "I am gambling that a curation of some of its past stories in their contexts, ancient and recent, will help alter current consciousness."[12] For Keller curation takes the form of "minding" and "dream reading" the "crowded, clouded, and prophetic metaphors" of the book of Revelation in which she puts them in relation to our precarious present in the hope that such juxtapositions might "release surprising relevancies."[13] Keller's complex curation of minding and dream reading is performative, not representational. It works with possibilities, potentialities that this curating of metaphors might actualize. It is this performative activism that Keller claims is needed for this critical moment, a "thought experiment of a theology not just multi-disciplinary but aiming through and beyond (trans) disciplinarity toward action."[14] Curating is a form of performative action that theology might find helpful in "assembling"—which includes both disassembling and reassembling—our collective futures amidst the political, ecological, and economic catastrophes with which we now live and breathe and have our being. As Walter Benjamin observes, "That things are 'status quo' *is* the catastrophe. It is not an ever-present possibility but what in each case is given."[15] To think these catastrophes theologically is to recognize their givenness, while curating them implies that they are more than given.

But curating? In comparison to the artist, the curator seems parasitic, serving only as a mediator, administrator, limited to merely selecting from the results of artists' creativity, a reality that has frustrated and infuriated many artists, like Robert Smithson, who bristled at the curator who "imposes his own limits on an art exhibit," a phenomenon he called "cultural confinement."[16] Moreover, art theorist Hal Foster cautions,

> Such curating suits the postindustrial aspects of an economy in which the appointed task for many people is to consume. . . . We cognitive laborers manipulate information, which is to say we curate the given, and this compiling often involves a great deal of complying.[17]

But the figure of the art curator is more than compiling and complying. David Levi Strauss suggests, "Curators have always been a curious mixture of bureaucrat and priest."[18] (I would add, also, a double shot of extrovert.) They are on one hand invisible. They experience the results of their work in exhibitions in art museums and galleries, yet their own creativity metastasizes into the exhibition itself, becoming indistinguishable from the assembled works on view. On the other hand, however, they can be highly visible in elite social gatherings, like cocktail parties with rich art collectors and patrons. The art curator embodies the contradictions that haunt the visual arts and the institutions that preserve and present them. Are they and their images of practice expressions of transformation and liberation, capable of speaking truth to power? Or are they merely instruments of power and wealth, employees of the court? Are they necessary for human and earthly flourishing or are they irrelevant, merely forms of decoration, distraction, indulgence that aestheticize oppressive and extractive structures? As priests of the art world, curators perform the liturgies, administer the sacraments, and offer the homilies that sustain faith in the visual arts as a viable means by which we may experience our lives individually and collectively as pulsing with what Paul calls the fruits of the spirit, faith, hope, and love. Or at least, this is my contention. For even if these priests are compromised, their liturgies and the images their practices generate might still be effective for theological thought.

AN EXHIBITION ANNOUNCEMENT AT MOMA

Not long before COVID required the closure of art museums in New York City, I received an email from the Museum of Modern Art announcing the opening of a new exhibition and inviting me to attend one of its special "preview" dates for members. MoMA played a crucial role in my intellectual and spiritual formation as an art historian and curator as a graduate student at Stony Brook from 1989 to 1991. Since then, the museum has insinuated itself into my life, becoming perhaps the most sacred of spaces, enriching my emotional, spiritual, and intellectual life, encouraging, challenging, inspiring, and consoling me through the ebbs and flows of my professional and personal journey. And yet, like churches, the other sacred spaces in my life, MoMA is entangled in and compromised by power and wealth, complicit with hegemonic structures of every type, white-male-hetero sovereignties, global capitalisms, imperialisms. In fact, like churches, art museums rely on the self-serving wealth of those who benefit from such violent and oppressive structures and systems. The legacy of the visual arts in what is often called the "European tradition" is dependent on those structures and systems, reliant on patrons, clients for the very existence of those works we travel such great distances to see.

And while MoMA had come to serve a liturgical and sacramental purpose in my life, I know that it—like every art museum—is the site for other liturgies and sacraments that celebrate and perpetuate extraction, exclusion, exploitation, which are administered by museum clergy—the director, development officer, and yes, the curators—to serve a global elite who regard art museums as their personal playground and art as the currency for their desperate attempts to purchase cultural and symbolic power, what sociologist Pierre Bourdieu called that which *cannot* be bought.

For the exhibition at MoMA of which I was sent an announcement, there were patrons and sponsors dinners; the artist offered talks and exhibition walkthroughs to a select few "stakeholders," and the curator, Dr. Paola Antonelli, an Italian curator and head of the architecture and design department, gave private tours to board members, collectors, and other VIPs who supported the exhibition financially. And the exhibition itself no doubt was leveraged for increased financial support in the future.

The historical legacy of the Museum of Modern Art, founded in 1929, reveals an almost evangelistic attempt to bring "modern art" to the broader public through its innovative educational programs. But the engine that drives it is from an assemblage of cocktail parties, board meetings, and martini lunches. In fact, MoMA was founded by three wives of industrialists: Lillie P. Bliss, Mary Quinn Sullivan, and Abby Aldrich Rockefeller. Kelly Crow, *Wall Street Journal* art journalist, has called the art world the world's longest continual cocktail party, which simply moves from New York City to Los Angeles, Berlin, Beijing, Paris, Aspen, Basel, or Miami Beach—depending on the season and the art fairs.

How can the works of art on view at MoMA, then, be anything other than icons of exclusion and oppression, relics of white-Euro-hetero-male sovereignty? How can art museums be anything but irrelevant at best, complicit at worst during this collaborative struggle for solidarity with all human and nonhuman creatures in an entangled cosmos, even while our struggle demands gestures of creative courage and imagination that are embodied in and through the results of artistic practices that we visit art museums to see? Does the complicit and compromised nature of the art museum render the works of art irrelevant or even an obstacle to our idealistic activism?

T. W. Adorno was haunted by this, in large part from the long shadow of Auschwitz. He begins his unfinished manuscript, *Aesthetic Theory*, with the following sobering observation. "It is self-evident that nothing concerning art is self-evident anymore, not its inner life, not its relation to the world, not even its right to exist." Written shortly before his death in 1969, these words seem even truer today not only for art but for those institutions, like art museums, that present it, especially while the world slowly begins to reopen from COVID.[19] As the world experiences and learns to live with political, economic, and ecological crises, it seems that the art world and the world of the visual arts and its dogmas and doctrines of "beauty" and "aesthetic experience" are not only powerless and irrelevant as a remedy but also complicit as a cause. Consequently, the structures of thinking influenced by art museums and the artifacts on their walls seem mighty useless, indeed, and perhaps even an abrogation of our responsibility as thinkers and activists.

This vulnerable, fraught, and compromised situation in which art and art museums find themselves is not unlike the one in which theology finds itself. Given the complicity of many theologies with underwriting the very catastrophes with which we now live, why bother with theology, with God-talk, or even God-adjacent talk at all, of creation, belief, human dominion, transcendence, redemption, and the final judgment? Can't we just follow Morton's advice and simply stop "retweeting agricultural-age monotheisms?"[20] I admit to being tempted to do so. But, in the last analysis, I am drawn to both theology and the visual arts, as well as the contexts in which they are performed, exhibited, curated. I am committed to art museums, artist studios, churches, and seminar rooms precisely because their irrelevance and impotence strike me as utterly relevant, or at the very least, *potentially* relevant for me from which to think and act responsibly in the midst of multiplying multiplicities of political, economic, and ecological catastrophe. Both art and theology can find strength in such weakness and irrelevance. What Keller observes about theology goes for the visual arts and art museums, as well. "Theology, unlike so many other disciplines, does not feel muted by its own irrelevance."[21] The art museum and other public spaces for the presentation of works of art are resources from which faith, hope, and love can be re-curated, not merely as an attitude, a personal piety, but as a form of political activism and responsible citizenship.

THE CONCEPT OF THE CURATORIAL

But is the extent of the curatorial merely a kind of priestly ritual, bureaucratic administration, and socialite small talk—deployed as a therapeutic soothing of existential angst through art? Timothy Morton and Catherine Keller did not think so, as they strained to find language that could accommodate the multiplicity and complexity they were experiencing at this most critical of moments. Is there a way to bring into alignment Morton's and Keller's use of the concept with a more robust understanding of the curatorial as a practice and a discourse that helps, engages, "faces" the world rather than facilitates an aesthetic retreat? Might it revitalize the political potential of our experience of art in public spaces? What follows is a closer look at the concept of the curatorial to confirm Morton's and Keller's creative intuition.

It is no secret that the concept of the curatorial has migrated from art museums into the broader cultural domain. As a *New York Times* article suggested in 2020, "Everyone's a curator now, from social media feeds and playlists to meals and vacations."[22] Again, Hal Foster is skeptical. He observes that

> curating promises a new kind of agency, but it might only deliver a heightened level of administration: if people can be reduced to skill sets and arguments to takeaways, then cultural interests can be packaged as curated consumption.[23]

Foster's critique of the curatorial is trenchant and apt. The dangers of the curatorial as a figure and concept abound. Yet its appropriation as an instrument and creative administrator of neoliberal aestheticism and quietism does not in any way exhaust its power or obscure its emancipatory potential. Indeed, theology might be especially equipped to retrieve and reveal it. For it is, to my mind, always on the lookout for lost causes, tarnished metaphors, abandoned concepts, and desecrated images, forgotten concepts. Against Foster, I suggest that the curatorial, as a practice of thinking, *does* indeed offer a potentially distinctive agency—one that is collective, liberative, restorative, and performative, which both Morton and Keller sense, even if they do not explore it in depth. The curatorial, to paraphrase Martin Luther on faith, is not what people think it is.[24]

The figure of the curator developed from the so-called *cura* tradition in ancient Roman literature and in the bureaucracy of the Roman Empire as caretakers of aqueducts and public supplies. It connotes, on one hand, being weighed down with anxieties and troubles, and on the other, it is an expression of concern for the welfare of another, conscientiousness, and devotedness.[25] The most complete expression of this ancient Roman and Latin tradition is a fable by the Roman poet Gaius Julius Hyginus about a mysterious quasi-goddess figure named Cura, who, after crossing a river, fashions a human from the mud she finds on the other side. It owes its notoriety in philosophical thought to Martin Heidegger, who quotes it in full in *Being and Time* and devotes considerable time discussing it in relation to *Da-sein* as "care."[26]

The curator cares for works of art through their preservation, presentation, and interpretation. But the curator expresses care not only for objects and artifacts, but for other human beings with whom she *shares* these cared-for works. It is this public, social, and relational dimension of the curatorial that is most crucial for theology. As art historian Thomas Crow has shown, the annual salon exhibitions of paintings and sculptures in Paris during the eighteenth and nineteenth centuries were responsible for generating the modern concept of a "public."[27]

The curated exhibition is a social event with political potential. Moreover, it is an event that generates relationships with others. As Edouard Manet declared in 1866, "To exhibit is to find allies for the struggle."[28] Swiss curator Hans Ulrich Obrist observes, "The whole curatorial thing has to do not only with exhibitions, it has a lot to do with bringing people together."[29] And curatorial theorist Jean-Paul Martinon puts it in more philosophical language: "The curatorial is the space of concern for the other."[30] The curatorial generates relationships, relationships among works of art, and then makes possible relationships with viewers. Martinon writes:

> As our bodies move in space, the curatorial proceeds by inventive steps or missteps from space to space. As such, the curatorial is a disruptive generosity . . . that can never be properly translated into language and always gives the slip to the economy of received and exchanged knowledge.[31]

It is this description of curating that possesses significant theological potential. What might a "disruptive generosity" look like in theological discourse and practice? How might a theological concept or idea stop us in our tracks, seduce us to come closer, to circle back through it for another look?

For Timothy Morton relation is all. To think ecologically is to think in relation, for it is "a radical openness to everything."[32] It is

> the thinking of interconnectedness. . . . It's a practice and a process of becoming fully aware of how human beings are connected with other beings—animal, vegetable, or mineral. Ultimately, this includes thinking about democracy.[33]

This "interconnectedness" threads together the ecological, political, and economic through human and nonhuman experiences and relationships.

> Ecology isn't just about global warming, recycling, and solar power—and also not just to do with everyday relationships between humans and nonhumans. It has to do with love, loss, despair, and compassion. It has to do with depression and psychosis. It has to do with capitalism and with what might exist after capitalism. It has to do with amazement, open-mindedness and wonder. . . . It has to do with coexistence.[34]

To experience art in art museums is not a solitary contemplative experience.[35] The art museum is a bustling—even chaotic—public square buzzing with multiplicities: bodies, discourses, experiences, and competing interests that represent the hive of energies that can be only loosely assembled, marshaled, guided by the decisions of a curator or a team of curators. The curatorial dimension of the art museum encourages experiences that are nonlinear, imaginative, and creative. It generates the potential for ideas, concepts, artifacts, and images in relation to provoke affects, feelings, emotions, memories, experiences, and concepts that defy "logic" and resist prediction (and mere repetition), description, and explanation. In an essay on Robert Rauschenberg's white paintings in 1951, the composer John Cage suggests, "'Art is the imitation of nature in her manner of operation.' Or a net."[36] The curatorial functions something like Cage's net—attracting, collecting not only works of art on view in a room but *all* that occurs therein, notwithstanding our own thoughts, feelings, memories as well as the bodies, the sounds, the movement within which our experience of the art takes place.[37] The curatorial casts a net, yet that net is neither constricting nor constant—its contents are always changing as we negotiate the galleries, moving through and experiencing works of art in always new ways, with new assemblages. The curatorial is spatial and material thinking and practicing in relation.

CURATION AT JUNCTURES

The exhibition at MoMA for which I was sent an announcement was an ambitious expression of curatorial activism. It featured the work of

Israeli-born artist-scientist-researcher Neri Oxman. "Her practice," Antonelli writes,

> is a powerful anticipation of a better possible future, and it is projected toward that future without apparent anxiety, patiently weaving connections between disciplines and between species, slowing down the pace of making by marrying the latest technologies with the most ancient and deliberate of tempos—those of silkworms, bees, and microbes. She engages change using change's own momentum.[38]

Oxman's work "explored natural and biological phenomena and the potential to use computation to reconstruct them at larger scales, demonstrating how this new technology could inform the future of designing and making objects." Oxman calls her practice "material ecology," which she describes as "an emerging field in design denoting informed relations between products, buildings, systems, and their environment . . . with emphasis on environmentally informed digital design and fabrication."[39]

Antonelli has devoted her curatorial career to exploring and presenting ways in which design might counter, slow, or reduce the catastrophic impact of the human animal on earth and its other fellow nonhuman inhabitants, a curatorial practice that she calls "restorative design." Oxman is a professor at MIT and founded the Mediated Matter Group in 2010, which is the center of her collaborative practice and includes computer scientists, designers, architects, biologists, a medical engineer, a mechanical engineer, an artist, a weaver, and a marine scientist.[40] This group is also under the aegis of MIT's Media Lab, which should raise some eyebrows. One of its important funders was Jeffrey Epstein, whose ubiquitous presence in the contemporary art world undermines the relevance of the contemporary art world to offer creative resources for the present moment. Leon Black, the ex-chairperson of MoMA's powerful board, is a Wall Street investor whose name adorns several galleries and is one of the world's foremost collectors of works on paper and has contributed significant funding to support the visual arts at Dartmouth College, including the newly renovated Hood Museum of Art.[41] Black also happens to have paid $130 million at auction for Edvard Munch's *The Scream* in 2013.

Last fall the director of the Media Lab, Joi Ito, was forced to resign amidst his relationship with Epstein. To participate in the visual arts, whether at

MoMA or at the Sheldon Museum of Art in Nebraska, where I served as chief curator for eleven years, is to be always already complicit, entangled with forces that conspire to snuff out the tiny flickering flame of prophetic possibility that the visual arts can—through great effort and cost—ignite. The irony, of course, is that far from suffocating it, these patrons and stakeholders claim that they are its most passionate and committed supporters.

Oxman's personal profile also might raise eyebrows.[42] Her "real-world boyfriend" is a "contrarian hedge-funder" who collaborates with Björk and who has attracted Brad Pitt's interest (brushing aside rumors that they were dating). She represents the kind of cross-over artist-personality of which Foster is deeply critical, offering a kind of cultural entertainment more suitable to the style section of the *New York Times*. Is this enough evidence to dismiss these projects as the kind of "cross-over" entertainment and celebrity culture that Foster criticizes?

This dark and problematic portrait is not unique to high-stakes art museums like MoMA. All art museums in one way or another are "ceremonial monuments" that "affirm the power and social authority of a patron class" and are thus a crucial part of "late capitalist rituals."[43] Art museums are hubs for the tripartite vectors of extractive and oppressive economics, ecology, and politics. Perhaps more than hubs, art museums perform a kind of cultural alchemy, transforming oppressive, nondemocratic, and extractive energy into the aesthetic realm of "art" and "culture."

But it is also in this hub, at the intersection of these destructive energies, that the curator works. Therefore, complicity, compromise, and bad faith always lurk. To be sure, many curators are content to participate unreflectively in this cultural alchemy. But the curator has at one's disposal resources to do otherwise. As Swiss curator Hans Ulrich Obrist observes, the curator is a powerful and necessary concept for responding to the "incredible proliferation of ideas, information, images" who more than "fills a space with objects," but "as the person who brings different cultural spheres into contact, invests new display features, and makes junctions that allow unexpected encounters and results."[44] It is these junctions that generate the unexpected that are crucial not only to the curatorial but, I argue, to the theological.

This means that the curator must work with people and systems with power and wealth to generate projects that might do more than merely serve to entrench the status quo. They might indeed be some of the "master's tools" that can never fully "dismantle the master's house," as Audre Lorde tells us. But perhaps the figure of the curator suggests that all such activism, especially theological activism, must be, in some way, an inside job that is always already implicated, compromised, tainted. The curator knows, feels, and must work with this tension. Does the theologian? Do theologians have the courage to risk compromise and complicity, work at that dangerous hub of wealth and power, hypocrisy and self-dealing, to realize their projects in ways that can be shared with multiple publics and be read and experienced in different ways that only *might* offer glimpses of hope?

For Obrist, quoted as the epigraph of this essay, curating "allows different elements to touch," which forms what he calls "junctures."[45] In order to bring a project to public view, the curator must touch and be touched by those with power and wealth, with competing visions and conflicting agendas. And it is through this curatorial activity—potentially compromising and risky at every turn—of assembling disparate and contradictory parts that the curator can offer an exhibition—cast a net—that allows other different elements to touch new junctures of thought, feeling, and experience that might catalyze, provoke new and unexpected forms of emancipatory thought and action. These "junctures" that the curatorial produce take place not only in the public realm of the experience of the exhibition, but backstage, in the storage facilities, boardrooms, staff meetings, lunches with potential patrons, and studio visits that bring the exhibition to visibility. These are often the relations that endure beyond particular projects, that continue to unfold new ways of thinking and acting collectively that may take other forms than exhibitions. In reflecting on my curatorial experience, the most important and memorable events are not visible, and no record of them exists. Theology, I suggest, should become more than the sum of its visible and legible parts—its books and journal articles and conference papers. Can theology, following the lead of the curatorial, consist of more than what is read? Is it possible to think theology curatorially by thinking about the relations, solidarities, partnerships, and experiences that cannot be pointed to in this or that text, but is everywhere present? Perhaps they might be foregrounded in different ways?[46]

Although implicated in the style section, tabloids, and boutique intellectual culture of *TED Talks* and perhaps an excessive faith in scientific innovation, Paola Antonelli's collaboration with Neri Oxman offers a serious contribution to assembling—curating—futures in the face of catastrophe (and often in collusion with those whose actions expedited its arrival). It is noteworthy that Antonelli points out Oxman's practice as devoid of anxiety and executed with patience. This attitude is certainly uncharacteristic of the apocalypticism that pervades the political left in regard to "climate change" as it performs its own version of a fundamentalist reading of the book of Revelation as the End of the World.

Since its founding, the Museum of Modern Art has considered architecture and design to be crucial for the transformation of modern life. This project represents such an effort. It makes visibly accessible and experientially present to an audience examples of Oxman's creative research into developing sustainable materials that collaborate with rather than control the environment, which is a direct response to the present moment. The designer's responsibility, Antonelli suggests, is

> to promote and enable synergy between the built and the grown, the artificial and the natural. . . . The designer then becomes a mediator, a gardener, an alchemist operating across scales and domains to conduct rather than construct.[47]

An exhibition, Obrist argues, can go beyond merely illustration or representation. It has the potential to produce—perform—reality, to realize something that did not exist before.[48] The potential inherent in curatorial thinking is that its value is located in the relationships of the constituent parts and its experiential, imaginative affect. Its assemblage of human and nonhuman relationships, both visible in the exhibition space and invisible, collaborated to bring the exhibition into being. In addition, it encourages the free circulation of human beings as physical, emotional, spiritual energies throughout the rooms that create something new, something different each time for the thousands and thousands of visitors who walk through the galleries. How might theological practice invite such free circulation?

There is a significant danger, however, that Oxman's artifacts become overly aestheticized in the galleries at MoMA, are dissociated from their original function as utilitarian materials, and are venerated as "Art" as they hang on the walls or are displayed on pedestals behind Plexiglas. And yet, this danger is well worth the risk, it seems to me. *Material Ecology* visualizes, offers images of what the future might be made of, presents the results of creativity and imagination that is occurring in the very boiler room of neoliberalism politics and economics. And yet, as this exhibition performs a possible future, it also solicits and invites new experiences and responses from artists, scientists, engineers, architects, activists, and even theologians, thus serving as a tool box from which to create new and other images, concepts, ideas, and artifacts—indeed, new curated ways of being and doing in the midst of this threefold catastrophe, but from which we must continue to honor life, our life and the lives of other human and nonhuman creatures. This is not the time for anorexic restriction from the aesthetic, from beauty, from feeling our bodies and our emotions that the curated experience of the arts facilitates. For Morton, it is the art's capacity to give access to beauty that is crucial for ecological thinking and acting because it, like catastrophes and other "hyperobjects," is ungraspable.[49] The creeps of the apocalypse will not be stopped, the various ecological, political, and economic catastrophes "solved" or "eliminated" (*or even understood*) through voting, protests, and governmental policies and by screaming and yelling that the End is near. The threefold catastrophe can, however, and needs to be curated creatively again and again. The curator moves on from project to project, from relationships to relationships, collaborators to collaborators, many of them less committed to healing the earth and the polis. As Obrist observes, "Curating, after all, produces ephemeral constellations that disappear."[50] These ephemeral constellations, like *Material Ecology*, pop up like a mushroom amidst the scarred and burned earth, and just as quickly return to it. Antonelli works with Oxman for several years on *Material Ecology*, and it is over in a matter of a few months. A new exhibition curated by a new team of curators takes its place. And Antonelli and Oxman go their ways pursuing new projects, new collaborations. Yet they are changed by that "ephemeral constellation," which continues to give and reward reflection, perhaps even increasing in value and relevance over time. If this sounds overly idealistic at this point, it probably is. It is a confession of faith.

TOWARD A CURATORIAL THEOLOGY

But where is the "theology" in all of this? How might putting exhibitions together—thinking and practicing the curatorial—meet this moment responsibly and prophetically? How might curatorial thinking generate new relationships with theological concepts, ideas, practices, and images that invite a free circulation that is as concerned with how it feels, how it is experienced, embracing chance, the unintended, and that privileges a nonlinear approach to discourse and practice? In addition, how might a curatorial theology lean into collaborations, into the unknown, the spontaneous, the unexpected, the unplanned gesture, the performance? How might a curated theology privilege experience, feeling, address the heart rather than the head, or rather than merely talking about the heart, talking about experience, do it, perform it?

It seems to me there is indeed a curatorial practice that is to be avoided. It is merely a selection and rearrangement of objects, concepts, images, and ideas that reflect the mechanism of the contemporary art world, with its relentless search for the new, the entertaining. But there are alternative expressions that emerge from a deep commitment and belief in the potential of art to transform lives, to create new collectivities and solidarities even amidst human and nonhuman suffering, oppression, and injustice. In short, it comes from a place of hope—and even its Pauline siblings, faith and love. The heart of the curatorial, I believe, *is* theological—that is, a concern for ultimacy, or what really matters, as an expression of and a search for faith, hope, and love—a gesture of what Deleuze called an attempt to restore belief in *this world*. A curator like Antonelli believes in Oxman's work and believes that by sharing it at MoMA it will transform her audience and those with whom she collaborated to bring it to fruition, in spite of the potential compromises and complicities she faced. Curators attempt to put the works of art, concepts, and images they believe in the best possible light to enable them to do their work in the world.

I might also suggest that the engine that drives creative, constructive, material, or radical theology is *curatorial*. Altizer called this "fundamental theology," which he claims "can only individually be enacted, and enacted through one's own thinking and one's own individual exploration."[51] Shaped by the (masculine) loners Nietzsche, Joyce, and Blake, Altizer fails here to recognize how his theology is shaped by relationships and a deep commitment to share it, to even perform it—to place in sometimes sharp and unexpected juxtapositions like Joyce and Augustine, Paul and Blake, Christ and Satan.

Like the art museum curator, the theologian is always already entangled in fraught and compromising relationships with power and wealth, in the church and academia, and working with a collection of images, ideas, and concepts that have been created for or appropriated as instruments of oppression, weapons of injustice, and sources for self-dealing. For the curator this recognition is part of one's work. At this moment, theology needs to recognize and navigate critically its own complicity. And yet, like the curator, the theologian must find ways to put these artifacts of thought into unexpected relationships, relationships that might yield new experiences of (not arguments for or representations of) God's presence in the world or the presence of hope, faith, and love. The theologian, thinking and practicing the curatorial, opens up the space *between* these ideas, concepts, and images—what Obrist called "junctures" or "zones of contact," which privilege and encourage multiple interpretations and experiences, empowering action, practice, activism.

The kinship between the theological and the curatorial is significant but largely unnoticed. Curatorial practice can help theologians deepen and expand their capacity to assemble *temporary* relationships among the irreconcilable multiplicities of ecology, economics, and political apocalypses with which we live that *curate* rather than interpret, explain, represent, solve. "If the curatorial can be seen to encourage something," Martinon claims,

> it will be to encourage artists, curators, and viewers to retain the vertigo that heightens the un-known: to be a warrior of the imaginary, abandoned to extremes and working without magnetic north or compass.[52]

I cannot imagine a better program statement for theology at this moment to empower theologians to curate risky exhibitions of ideas, concepts, artifacts, and images that might unlock the transformative potential of experiences generated by curating these multiple catastrophes theologically.

NOTES

1. Catherine Keller, "Creeps of the Apocalypse: Climate, Capital, Democracy," in this volume.
2. Isabelle Stengers, *In Catastrophic Times: Resisting the Coming Barbarism*, trans. Andrew Goffey (London: Open University Press, 2015), 20.
3. Timothy Morton, *The Ecological Thought* (Cambridge, Mass.: Harvard University Press, 2012), 8.
4. Morton, *Ecological Thought*, 10.

5. Morton, *Ecological Thought*, 11.
6. Timothy Morton, *All Art Is Ecological* (London: Penguin Classics, 2021), 28.
7. See Luc Boltanski and Eve Chiapello, *The New Spirit of Capitalism*, trans. Gregory Elliott (New York: Verso, 2018). The concept of the "creative self" plays a crucial role in this "new spirit of capitalism," as exemplified by the work of such creativity gurus as Seth Godin. See also Ann Jefferson, *Genius in France: An Idea and Its Uses* (Princeton: Princeton University Press, 2014) for a study on the role of artistic genius in modernity. See also Morton, *All Art Is Ecological*.
8. Morton, *All Art Is Ecological*, 78.
9. See Jean-Paul Martinon, ed., *The Curatorial: A Philosophy of Curating* (New York: Bloomsbury, 2013).
10. Timothy Morton, "Queer Green Sex Toys," in *Meaningful Flesh: Reflections on Religion and Nature for a Queer Planet*, ed. Whitney A. Bauman (Rome: Punctum Press, 2018).
11. Catherine Keller, *Facing the Apocalypse: Climate, Democracy, and Other Last Chances* (Maryknoll, N.Y.: Orbis, 2021), xii–xiii.
12. Keller, *Facing the Apocalypse*, xiii.
13. Keller, *Facing the Apocalypse*, xiv.
14. Keller, "Creeps of the Apocalypse."
15. Walter Benjamin, *The Arcades Project*, trans. Howard Eiland and Kevin McLaughlin (Cambridge, Mass.: Belknap Press of Harvard University Press, 1999), 473.
16. Robert Smithson, "Cultural Confinement," in *Robert Smithson: The Collected Writings*, ed. Jack Flam (Oakland: University of California Press, 1996), 154.
17. Hal Foster, "Exhibitionists," in *What Comes after Farce? Art and Criticism at a Time of Debacle* (New York: Verso, 2020), 68.
18. David Levi Strauss, "The Bias of the World: Curating after Szeemann & Hopps," in *Cautionary Tales: Critical Curating*, ed. Steven Rand and Heather Kouris (New York: Apex Art, 2007).
19. T. W. Adorno, *Aesthetic Theory*, trans. Robert Hullot-Kantor (Minneapolis: University of Minnesota Press, 1995). See David Joselit, "Art Museums Will Never Be the Same. That's a Good Thing," *MIT Press Reader*, July 22, 2020, https://thereader.mitpress.mit.edu/art-museums-will-never-be-the-same-thats-a-good-thing/.
20. Timothy Morton, "How to Defeat Invisible Gods," June 26, 2015, Mexico City.
21. Catherine Keller, "Letter to a Young Theologian," in *Letters to a Young Theologian*, ed. Henco Van Der Westhuizen (Minneapolis: Fortress Press, 2021).
22. Lou Stoppard, "Everyone's a Curator Now," *New York Times*, March 3, 2020.
23. Hal Foster, "The Exhibitionists," in *What Comes after Farce?*, 68.

24. Martin Luther, "An Introduction to St. Paul's Letter to the Romans" (1522).
25. See Stefan Nowotny, "The Curator Crosses the River: A Fabulation," in Martinon, *The Curatorial*. See also Maria Puig de la Bellacasa, *Matters of Care: Speculative Ethics in More Than Human Worlds* (Minneapolis: University of Minnesota Press, 2017); John T. Hamilton, "Homo Curans," in *Security: Politics, Humanity, and the Philology of Care* (Princeton: Princeton University Press, 2013). Hans Blumenberg explores Heidegger's use of the fable of Hyginus (64 BCE–17 CE), in *Care Crosses the River*, trans. Paul Fleming (Stanford, Calif.: Stanford University Press, 2010).
26. Martin Heidegger, *Being and Time*, trans. Joan Stambaugh, foreword Dennis J. Schmidt (Albany: SUNY Press, 2010), 184–86.
27. Thomas Crow, *Painters and Public Life in Eighteenth-Century Paris* (New Haven: Yale University Press, 1985).
28. Linda Nochlin, ed., *Realism and Tradition in Art, 1848–1900: Sources and Documents* (Upper Saddle River, N.J.: Prentice Hall, 1966), 81.
29. Hans Ulrich Obrist, *Everything You Always Wanted to Know about Curating* (Berlin: Sternberg Press, 2011), 34.
30. Jean-Paul, Martinon, "Theses on the Philosophy of Curating, in Martinon, *Curatorial*, 27.
31. Martinon, "Theses," 29.
32. Morton, *Ecological Thought*, 15.
33. Morton, *Ecological Thought*, 7.
34. Morton, *Ecological Thought*, 2.
35. We rarely encounter works of art in silence and solitude. Yet it is such constructed images—generated by cinema and photography—that create frustration for viewers of art exhibitions. This fantasy image encourages us to ignore or view as a distraction much of what is occurring to and with us as we experience art.
36. John Cage, "On Robert Rauschenberg, Artist, and His Work," in *Silence: Lectures and Writings* (Middletown, Conn.: Wesleyan University Press, 1961), 100.
37. In my personal experience of returning to art museums as New York City reopened after the COVID lockdown, I not only felt relief at seeing works of art I'd not seen for over a year, but I realized I'd missed the social and relational dimension of the art museum.
38. Paula Antonelli, "The Natural Evolution of Architecture," in *Neri Oxman: Material Ecology* (New York: MoMA, 2020), 13.
39. Antonelli, "Natural Evolution," 13.
40. Antonelli, "Natural Evolution," 15.
41. Leon Black was forced to resign as chair of the MoMA board in the summer of 2021 and not stand for re-election as a result of his relationship with Epstein;

see Robin Pogrebin and Matthew Goldstein, "Leon Black to Step Down as MoMA Chairman," *New York Times*, March 26, 2021.

42. Penelope Green, "Who Is Neri Oxman?," *New York Times*, October 6, 2018.
43. Carol Duncan and Alan Wallach, "The Museum of Modern Art as Late Capitalist Ritual: An Iconographic Analysis," *Marxist Perspectives* 4 (1978). The literature on the art museum and its relationship to coloniality and imperialism is vast. See also Rosalind Krauss, "The Cultural Logic of the Late Capitalist Museum," *October* 54 (Autumn 1990).
44. Hans Ulrich Obrist, "The Curator," in *Sharp Tongues, Loose lips, Open Eyes, Ears to the Ground*, ed. April Lamm (Berlin: Sternberg, 2014), 24.
45. Hans Ulrich Obrist, *Ways of Curating* (London: Faber & Faber, 2014).
46. Karen Bray, *Grave Attending: A Political Theology for the Unredeemed* (New York: Fordham University Press, 2019).
47. Antonelli, "Natural Evolution," 37.
48. Obrist, *Ways of Curating*, 167–68.
49. Morton, *Ecological Thought*, 4–5b.
50. Obrist, *Everything You Always Wanted to Know about Curating*, 51.
51. Thomas J. J. Altizer, *The Call to Radical Theology*, ed. Lissa McCullough (Albany: SUNY Press, 2012), 160.

❧ The Costs of Citizenship: *Politeuma* in the Letter to the Philippians

JENNIFER QUIGLEY

But our politeuma *is in heaven, and we are waiting for a savior from there.*
—PHILIPPIANS 3:20

INTRODUCTION

Unstable pasts and presents draw attention to the instability of citizenship. When I first wrote for a conference planned in early 2020, the local crises of democracy that lingered in mind included court cases in Georgia, Wisconsin, and Ohio,[1] where, in the name of good governance, conservative groups and administrations made legal appeals to purge voters from the rolls for not voting recently or not responding to mass mailings. Globally, I had in mind India, where the publication of the National Register of Citizens, in conjunction with the Citizenship Amendment Bill, left some two million mostly Muslim Indians with a limited window in which to prove their citizenship.[2] I also was thinking of how the erasure of Rohingya Muslims in Myanmar from the census beginning in 2014 alongside government and military refusal to recognize their citizenship undergirded the violence and ethnic cleansing that began in 2017.[3] Here, intertwined, we see the ways in which citizenship can be destabilized: being counted, legal redress, the ability to live in one's space, claims over land, history, identity, and, within democratic systems, a voice and vote in governance.

The now-past has given way to an even more destabilized present. A global pandemic in which the common good of public health has been polarized by government officials for personal political gain, and vaccine access reflects broader global economic and health inequalities. Persistent false claims of "voter fraud," culminating in a deadly insurrection at the United States Capitol building in the wake of the 2020 presidential election, have led to present and persistent legislative attempts to make it more difficult to vote. Our present contexts shape the questions historians bring to often fragmentary and local evidence from the past. Both writing about the past and the attention that ancient writings bring to our present help to imagine, and perhaps to assemble, possible futures.

At a time of citizenship instability in the first century CE, in a marshy and backwater Roman colony, a community of impoverished Christ followers sent letters and support back and forth with a colleague who was in jail.[4] In one of these letters, they were asked to imagine, in their endless waiting—for a letter, for Epaphroditus, for a savior—a present and a future in which they held *politeuma*, citizenship, in heaven. What does it mean to invoke, to assemble, a future of heavenly *politeuma*? What does it mean, anyway, to have citizenship in heaven? This chapter considers these questions and Paul's claim in the Letter to the Philippians that Christ communities have "citizenship (*politeuma*) in heaven." This is a singular *politeuma*; it is the only occurrence of this term in the letters cowritten by Paul.

Biblical texts, including the letters of Paul, continue to influence communities of interpretation, ethical debates, and political discourse. Paul has proven especially enticing for political, philosophical, and theological arguments about identity, belonging, and democracy; this first-century Jew-in-Christ has been invoked in recent philosophical debates over the foundation(s) of universalism (cf. Alain Badiou and Troels Engberg-Pedersen).[5] Here in Philippians, though, we have an argument about particularity, a claim to community membership and participation in a heavenly polity.

In this chapter, I argue that it is worth considering why this singular *politeuma* appears not in, for example, the Corinthian correspondence, where the *ekklēsia*, the democratic assembly, is the primary organizing principle of in-Christ community. Rather, citizenship is evoked in the Letter to the Philippians, in which *koinōnia*, or venture, is the organizing principle of in-Christ community. Given the unstable and complex legal and financial context of "citizenship" under the Roman empire in the first century CE and the prominence of what I call theo-economics in this particular letter, *politeuma* for Paul has much more to do with divine financial obligations and benefits than with self-determination or voting rights. In short, heavenly citizenship is more about taxation than political taxonomy. Theology and the economy are deeply intertwined in the imaginaries of this ancient Christ community as they assembled in person and across the distance through their correspondence with their colleague in Christ, the imprisoned Paul, and as they imagined possible futures while making do in an unstable present.

CITIZENSHIP AND COLONIZATION

To assemble a heavenly *politeuma*, it is helpful first to think through what *politeuma* is not.[6] *Politeuma* is not about a hoped-for future self-determination through voter participation; that would have been almost unimaginable for this community of mostly Greeks who had experienced intergenerational loss of land to centuriation and (de)citizenship through military occupation. The local historical context and occasion of the Letter to the Philippians give clues to this.

Philippi, situated in Northern Greece, lay along the Via Egnatia, the main east-west thoroughfare of the Roman Empire. With gold and silver mines in Mount Pangaiaon and proximity to the port of Neapolis, Philippi was a city whose resources and land were frequently stripped for imperial projects, with a long history of colonization and recolonization. A hundred years or so before Paul and the Philippian Christ followers exchanged letters, the Roman civil war landed on Philippi's doorstep. In 42 BCE, following the assassination of Julius Caesar by Brutus and Cassius, Rome fell into civil war, and Octavian and Mark Antony met Brutus and Cassius at Philippi. More than 200,000 troops descended on the city and devastated the local food supply, the land, and the city. Antony and Octavian won, and to celebrate their victory, they immediately colonized Philippi, releasing hundreds of their soldiers from service and handing the city over to them. Over the next fifteen years Philippi was renamed twice and received two more influxes of colonists, once when Octavian became emperor in 30 BCE and again when he received the title "Augustus" from the Roman Senate in 27 BCE. At this latter date the city was named *Colonia Iulia Augusta Philippensis*. Three times in twenty years, the emerging Roman imperial project enacted centuriation on the city of Philippi.[7] Coins were minted depicting these colonizations. The newly minted emperor newly minted coins to commemorate military victory and land distribution to veteran-citizens.

These often depicted either military trophies and/or a man leading oxen at plowing to represent the process of dividing land through centuriation.⁸ The members of the Christ communities at Philippi were likely descendants of those who had lost land in this reshuffling of political and economic powers.

Colonization, conquest, and unstable citizenship go hand in hand under empire. Citizenship was rapidly expanding in the first century CE. By the time Paul is writing from prison, the Roman imperial project has been in full force for a century. The narrative of Philippi is hardly unusual for this period. To aid in military conquest and imperial expansion, emperors would incentivize military service with citizenship, centuriation, and colonization. At the time of these first colonizations, Philippi was the only known overseas colony, but a hundred years later, this was no longer the case; nearly every province of the empire had an official *colonia*, largely (re)populated through centuriation.⁹ There are far more Roman citizens in the mid-first century CE than in the mid-first century BCE.

But citizenship in the mid-first century CE is also quite unstable. Veterans in the provinces may have had citizenship, but the rights of Roman citizens were, unsurprisingly, not equal, and in fact, eroding. During the Augustan age, very little evidence exists that provincial citizens could participate in the *ius honorum*, the right to seek office in Rome; according to Sherwin-White, provincials were not recognized by elections officials or granted the *latus clavus*, which indicated imperial approval of candidacy.¹⁰ These citizens were too new to count. To take a single example, the *tabula clesiana* of 46 CE from Caesar Claudius grants citizenship to residents near the Alps and suggests that those who demonstrate "loyalty and wealth" may aspire to the senate. At the same time, Claudius only acknowledges three names who might possibly stand for an election; almost no citizen seems to be quite loyal or wealthy enough to count for full citizenship rights.¹¹ As Rome shifts from a republic to empire, access to governmental self-determination and office of its citizens is reduced. And for new citizens, these possibilities are even more limited.

I am not suggesting that the Philippian Christ followers knew the fine-grain details of elite citizenship and its increasing instability—precisely the opposite. The Philippian Christ followers were members of the 99 percent. Rather, I am arguing that the ancient political imaginary for self-determination was limited for all inhabitants of the Roman empire and even more so by the time Paul finds himself in jail. Instead, we might think of the *ius latinum*, an emerging middle legal state at this time period, in which some persons who were not quite citizens but not quite not citizens had basic legal rights under Roman law. These included the right of *commercium*, to enter into business contracts, including acquiring and inheriting property, as well as the right for legal redress for such contracts and access to basic legal due process.[12] Early Christ followers in Philippi would have understood such contractual and legal possibilities as held by some of their neighbors and perhaps accessible to them, if not in the present, then in the future.

ASSEMBLING BODIES IN CORINTH AND PHILIPPI

The term that Christian communities translate as "church," *ekklēsia*, is found throughout the letters of Paul. *Ekklēsia*, a political term, is the language of democratic assembly. Many early Christ communities used the term *ekklēsia* to organize themselves and communicate something about their community identity; the Corinthian communities are an excellent example of this. Anna Miller's book *Corinthian Democracy: Democratic Discourse in 1 Corinthians* explores the political and theological implications of *ekklēsia* discourse in 1 Corinthians:

> Paul's rhetorical aim . . . is marked by rhetorical tactics drawn from a robust discourse of democracy—what I term *"ekklēsia* discourse"—a discourse pervasive in the eastern Roman Empire and within the Corinthian community. . . . Paul's rhetoric is inscribed with this *ekklēsia* discourse in order to make his own leadership legitimate in a context where the same discourse was being mobilized to construct a community around the actions of empowered, free citizens. . . . [I] trace a persuasive ancient discourse of democracy that emerged out of the Greek civic institution composed of the body of free citizens, the *ekklēsia* . . . this ancient Christian community was constructed, at least rhetorically, as a civic and political body . . . disparate community members at Corinth took a vocal role in deliberative decision making within this Christian community.[13]

Ekklēsia discourse, when deployed by Christ followers in Corinth, marks a community as a civic and political body. The Corinthians already understand themselves as having self-governance through the model of the *ekklēsia*. *Politeuma*, though, does not occur in the Corinthian correspondence, but in a letter to the Philippians. How do the Philippian Christ communities describe and organize themselves, and how, then, does *politeuma* fit into this model?

I would argue that *koinōnia* is the principal way in which the Philippians describe their community. It is difficult to underestimate the importance of Paul's use of *koinōnia* language throughout the letter.[14] The term appears six times in the very brief letter, compared with only two occurrences of *ekklēsia*;[15] in Philippians, Paul uses *koinōnia* discourse instead of *ekklēsia* discourse. At a time before Christ communities had fully settled into the language they would use in self-description, what does it mean to invoke a *koinōnia* for a community?

Recent work by Julien Ogereau has demonstrated that *koinōnia* is primarily a term with financial valences. *Koinōnia*-cognates "essentially expressed the idea of partnership, be it economic, political, marital, or otherwise."[16] Despite its frequent translation as "sharing" or "fellowship," its nearly ubiquitous use comes from contracts, from land leases to marriage contracts to camel sales. It might better be translated as a "venture," one in which mutual risk and reward are shared.

If two people were renting land to grow crops, they would likely draw up a *koinōnia*. They would agree to split the costs of everything from rent to seeds, they would share the labor of tilling the soil, of planting and watering, and weeding, and, at the right time, they would share the reward of the harvest together. There is shared risk and shared reward in farming. The crop might fail, insects might come, the rains might never fall. By working in a *koinōnia*, those risks and losses are spread out so as not to hurt as much, and the rewards of a plentiful harvest are also shared.

This is precisely the imagery Paul uses in greeting the Philippians, with whom he shares a *koinōnia* in the gospel. Philippians 1:5–7 reads a little differently when attention is paid to the financial valences of the text, especially to the *koinōnia* that the Philippian communities participate in with Paul.

> [I give thanks] . . . because of your venture (*koinōnia*) in the gospel from the first day until now. I have confidence in this, that the one who began a good work in you will complete it by the day of Jesus

Christ. Even as it is right for me to think this way about all of you because I have you in my heart, both in my imprisonment and in my defense and warranty of the gospel, since all of you are my joint-shareholders in grace.

The Philippians are asked, after all, to yield a "harvest of righteousness" (1:11). The Christ communities in Philippi, with Paul, have assembled themselves as a mutually invested and financially bound body. This makes sense, given the situation of the letter. Paul, writing from prison, has been receiving support, including financial support, from the Philippians, brought by Epaphroditus, who might have been an enslaved person.[17] If the Philippians and Paul have been sharing the risk and the reward of the gospel venture together, Paul, by landing in jail, might have caused harm to his partners and perhaps even to the gospel. Paul writes to justify that his imprisonment is actually a boon to their shared work in the gospel and that they should not worry about their *koinōnia*, despite evidence to the contrary.

This is likely not the first or the last time that Paul has taken money or will take money from the Philippians. Paul acknowledges in Philippians 4:15 that in the early days of the gospel, no other Christ assembly *koinōnia*-ed with Paul except them. In 2 Corinthians 8, Paul uses the Philippians' generosity to guilt the likely wealthier Corinthian assemblies for not sharing as much from their resources. And Paul seems to hope and expect throughout Philippians that this community will continue to contribute to (his and) the communities' needs, no matter the circumstances. The text in which we find the imagery of a *politeuma* in heaven has, at its core, economic themes. This Philippian body, assembled as a *koinōnia* in the gospel, understands their work in terms of mutual aid and investment, of shared risk and reward.

THEO-ECONOMICS IN THE LETTER TO THE PHILIPPIANS

These themes, about money, prison, and mutual investment in community, are not only financial; they are also theological. This is unsurprising, given that we have significant evidence that our modern scholarly categories of theology and economics were not so clearly delineated and separated in antiquity.[18] Many texts, documents, and objects instead demonstrate what I call theo-economics,[19] or an intertwined theological and economic logic in which divine and human beings regularly enter into transactions with one another. Accounting for god(s) and an accounting of the ordered world, *theologia* and *oikonomia*, are inseparable in the ancient world.[20] We have significant evidence in antiquity that divine and semi-divine beings were understood as having vibrant materiality within the economic sphere and that the gods were economic actors with whom humans could transact.[21] When a goddess has her own bank account, when a god stands to lose his property unless an emperor intervenes, and when divine and semi-divine beings receive dedicatory tithes, enter into loan agreements, and create and regulate their own currency, modern scholars need a different framework than "metaphor" to account for these phenomena. Such a framework must take seriously the ancient worldview that divine activity in the economy is not only possible, but quite normal.

When we turn to theo-economic language in New Testament and early Christian texts, then, we need to understand such language in its broader context. These texts emerge in a context in which we find divine activity in human economics, a divine economy with its own commodities and transactions, and the ability of humans to effect activity within that divine economy. Paul's Letter to the Philippians contains texts such as, "For me, living is Christ and dying is gain" (1:21); "I calculate all things as a loss . . . so that I might gain the profit Christ" (3:8); and "And my God will fulfill all your lack according to his riches in Christ Jesus" (4:19). Such language participates in ordinary divine-human theo-economic logics that would have been legible to early Christ followers. They saw and experienced divine involvement in the economic sphere both among others and in their own communities.

A HEAVENLY POLITEUMA

Perhaps a *politeuma* in heaven can also be read through the lens of theo-economics. After all, while it is clear that *politeuma* in the mid-first century CE cannot be about voting rights or self-determination, it does invoke rights with legal and financial implications. A significant portion of the *ius latinum* is related to contracts and court access through the *commercium*.

Citizenship also has significant tax implications. Roman citizens were almost entirely exempt from paying taxes in the imperial period. While there were various ways wealthy citizens in the provinces could be taxed on their wealth, Roman citizens were exempt from the *tributa*, taxes that were levied upon noncitizens based upon census data.²² Noncitizens who crossed the empire could also find themselves subject to additional punitive taxes. A well-known example is the *fiscus judaicus* that Vespasian levied on Jews following the destruction of the temple in Jerusalem, in which the funds were designated for the temple to Jupiter in Rome.²³

On the other hand, citizens also had access to what few social safety nets existed in antiquity. In the ancient city-state system, citizenship status was often required to receive benefits from the system of euergetism.²⁴ For citizens in Rome, this meant access to the well-known *cura annonae*, or the grain dole, a distribution that began in the second century BCE as subsidized prices for grain for Roman male citizens, but that by the end of the first century BCE had morphed into a free monthly grain distribution for all Roman citizens in the city over the age of ten; the amount was likely more than was even needed to feed a single person for a month.²⁵ To supply the grain dole for the some 150,000–200,000 citizens required a complex and extractive system of cultivation and shipping that relied heavily on the labor of noncitizens in the provinces, especially in Egypt. Noncitizens throughout the empire knew about and participated in the process of feeding Roman citizens.²⁶

Paul, then, is invoking a theo-economic imaginary in which the Philippians have access to a divine economy. The range of possibilities for participation in a heavenly *politeuma* might include exemption from divine taxes, legal recognition of the financial relationships they have had with one another and with their fellow Christ-followers, or a system of divine legal redress for financial abuses they may have encountered. Perhaps most importantly, this theo-economic imaginary might have included access to a divinely distributed social safety net, an ancient universal basic income that offered enough divine provision to nourish the entirety of the Christ community. For Paul, who writes in this letter that he has known hunger (Philippians 4:11–12), and for the members of the community who likely had experience with food insecurity, this is a theo-economic imaginary whose apocalyptic pangs include hunger.²⁷

For a community that largely would have lacked access to these benefits within the Roman empire, such an imaginary, even constrained by its imperial model, is radical. For folks who would have been more used to paying in for others' citizenship benefits rather than enjoying them themselves, imagining their own future financial rights while scraping together enough to help their imprisoned friend assembles a theological economy of abundance even in the midst of lack. For this community, which found enough to go around even when there was not enough to go around, and for this community that stretches itself to send support to a colleague jailed by the legal system of empire, to imagine a heavenly *politeuma* from which a savior might come is both a contextually bound and radical act. After all, the "The Christians' governing institution is outside Philippi: in fact, outside the Roman empire."[28] For these Christ followers who had already organized themselves into a *koinōnia* of mutual support, they were beginning to assemble a heavenly polity. For this community in Philippi that had experienced the costs of citizenship, from land loss to taxes to military occupation, to begin to imagine its limited benefits, provided by God, was a significant theological and economic shift.

CONCLUSION

it was clear they were hungry
with their carts empty the clothes inside their empty hands . . .
the asphalt street on the red dirt the dirt taxpayers pay for
up to that invisible line visible thick white paint
we didn't know how they had ended up that way
on *that* side
we didn't know how we had ended up here
we didn't know but we understood why they walk
the opposite direction to buy food on this side
this side we all know is hunger[29]

Salvadoran poet Javier Zamora wrote the poem "Citizenship" to describe his experience watching homeless U.S. citizens cross the port of entry in Nogales, Arizona, to buy cheaper food in Mexico. It raises poignant questions about the modern conception of citizenship, both in terms of how it contributes to human flourishing (or fails to) and how citizenship often constructs boundaries that need to be crossed. Citizenship is a shifting and unstable category with economic, political, and, at least in Christianity, theological entanglements. In this chapter, I have explored some of these entanglements by considering the invocation by the first-century Christ follower Paul to an impoverished Christ community in a marshy and backwater Roman colony that these community members were participants in a heavenly *politeuma*. I began by demonstrating that using *politeuma* in the mid-first century CE to suggest voting rights or self-determination would have been illegible. The rapid expansion of the Roman imperial project extended citizenship while simultaneously gutting many of its rights for those living outside of Rome. Instead, in a space and community that had experienced violent military action and centuriation, imagining citizenship was to imagine basic legal rights to enter contracts, seek basic legal redress, and perhaps to hope for access to some basic social safety nets, including supplemental nutrition. These aspects of citizenship are also worth considering, given the situation of the Letter to the Philippians, a text in which finances are a central theme.

A comparison between *ekklēsia* in 1 Corinthians and *koinōnia* in Philippians revealed the ways in which the Christ community in Philippi has assembled itself as a financially bound and mutually invested body. Attention to the local, occasional, and contextual aspects of the letters of Paul and the people beside him perhaps has as much if not more to offer than the universalist Paul imagined by contemporary philosophy.[30] Transdisciplinary work only benefits from such particularity, from attention to difference and to small but significant details such as tax practices and grain doles. Paul and the Philippian community lived, worked, wrote, read, and listened "ignorant of the future," as must we all.[31] After all, what small bits of our present will historians assemble in(to) our future in their writing of the past?

I have also considered the ways in which theology and economics are intertwined in antiquity; we should imagine a theo-economic system in which gods and humans regularly transacted with one another. To imagine, then, that a heavenly citizenship with a divine savior figure has economic implications should be unsurprising. Lastly, I considered the imaginative theo-economic possibilities that the Philippian assemblies might have hoped for: divine tax breaks and exemptions from the extractive tax system they experienced as noncitizens of Rome, divine legal recognition of their contractual relationships with one another, which might have no redress in the earthly Roman legal system, and perhaps access to a divinely provided supplemental nutritional program in a community that knew hunger. While the invocation of a heavenly *politeuma* and a divine savior figure might have been limited by the contextual constraints of living under empire, the Philippians and Paul were beginning to imaginatively assemble a potential future with an alternative economic theology without financial/legal exploitation and with divine abundance. The Philippians were already attempting to live this possible future in the present through the mutual aid and investment of a *koinōnia* in the gospel.

NOTES

1. Reis Thebault and Hannah Knowles, "Georgia Purged 309,000 Voters from Its Rolls. It's the Second State to Make Cuts in Less than a Week," *Washington Post*, December 17, 2019, https://www.washingtonpost.com/nation/2019/12/17/georgia-purged-voters-its-rolls-its-second-state-make-cuts-less-than-week/; German Lopez, "Supreme Court's Conservative Justices Uphold Ohio's Voter Purge System," updated June 11, 2018, https://www.vox.com/policy-and-politics/2018/6/11/17448742/ohio-voter-purge-supreme-court-ruling.
2. Sigal Samuel, "India Just Redefined Its Citizenship Criteria to Exclude Muslims," December 12, 2019, https://www.vox.com/future-perfect/2019/12/12/21010975/india-muslim-citizenship-bill-national-register.
3. "Myanmar Rohingya: What You Need to Know about the Crisis," *BBC News*, https://www.bbc.com/news/world-asia-41566561.
4. For more on the prison correspondence of Paul, see Ryan Schellenberg, *Abject Joy: Paul, Prison, and the Art of Making Do* (New York: Oxford University Press, 2021).
5. Alain Badiou, *St. Paul: The Foundation of Universalism*, trans. Ray Brassier (Stanford, Calif.: Stanford University Press, 2003); Troels Engberg-Pederson, "Paul and Universalism," in *Paul and the Philosophers*, ed. Ward Blanton and Hent De Vries (New York: Fordham University Press, 2013).
6. Although it is beyond the focus of this essay, some scholars have contended that *politeuma* is invoked in this polemical passage to contrast with other Jews in Philippi. Peter Oakes has recently convincingly argued that an examination of Jewish papyri that use *politeuma* use the term not to invoke formal ethnic identity but rather to discuss structures and processes of governance, including legal redress; Oakes, "The Christians and their Politeuma in Heaven: Philippians 3:20 and the Herakleopolis Papyri," in *In the Crucible of Empire: The Impact of Roman Citizenship upon Greeks, Jews, and Christians*, ed. Katell Berthelot and Jonathan J. Price (Leuven: Peeters, 2019).
7. For more on the archaeology of Philippi, see Jennifer Quigley, "Philippi," http://www.oxfordbiblicalstudies.com/article/opr/t998/e53.

8. See, for example, Roman Provincial Coinage (RPC), 1656. For more on Roman centuriation, see Laura Nasrallah, *Archaeology and the Letters of Paul* (Oxford: Oxford University Press, 2019), 118–22, and Lawrence Keppie, "Colonisation and Veteran Settlement in the 1st century AD" ("Colonisation and Veteran Settlement in Italy in the First Century A.D," on *JSTOR* [1984]).
9. A. N. Sherwin-White, *The Roman Citizenship* (Oxford: Clarendon Press, 1973), 230. While scholars debate the precise differences among terms, it is worth noting that east of Rome, while there are numerous *coloniae* established, including Philippi, only Stobi is designated as an *oppidum civium Romanorum* and not *municipia*. *Municipiae* were cities outside of Rome whose land was legally designated as Rome (229). During the time of Vespasian in the last quarter of the first century CE and continuing through the rest of the Flavian dynasty, many more cities were designated as *municipiae*, and citizenship was extended to all residents of Spain (360–65). Not coincidentally, this occurred at the same time as other imperial subjects, including those who revolted in Judea, experienced increasing violence.
10. Sherwin-White, *Roman Citizenship*, 234–36.
11. Sherwin-White, *Roman Citizenship*, 239–40.
12. George Long, "Latinitas," in William Smith, *A Dictionary of Greek and Roman Antiquities*, http://penelope.uchicago.edu/Thayer/E/Roman/Texts/secondary/SMIGRA*/Latinitas.html.
13. Anna C. Miller, *Corinthian Democracy: Democratic Discourse in 1 Corinthians* (Eugene, Ore.: Wipf and Stock, 2015), 1–2.
14. Mark A. Jennings has also recently noted the centrality of *koinōnia* as an organizing principle in the letter, arguing that "Paul does not exhort the Philippians to be partners in the advance of the gospel as generally conceived, but rather, that they be his partners in his gospel mission"; Jennings, *The Price of Partnership in the Letter to the Philippians* (New York: Bloomsbury T & T Clark, 2018), 25.
15. The three-to-one *koinōnia*-to-*ekklēsia* ratio found in Philippians might not seem significantly imbalanced at first. However, the context of the two appearances of *ekklēsia* helps to demonstrate that the term is supplanted by Paul's deployment of *koinōnia* cognates. *Ekklēsia* occurs at Philippians 3:6 and Philippians 4:15. In Philippians 3:6, in the midst of an autobiographical section in which Paul talks about his reasons for "confidence in the flesh" (Phil 3:4), he adds, "according to zeal, a persecutor of the *ekklēsia*." Here, Paul's pursuit of the assembly is buried in a long list that includes Paul's status markers within Judaism. This first mention is in reference to Paul's past and has no immediate connection to the Philippian community or its context. The

second appearance, in Philippians 4:15, occurs in a passage about Paul receiving financial support from the Philippians. There he writes, "No assembly invested with me in the matter of giving and receiving except you." Here, while Paul acknowledges that the assemblies with which he corresponds commonly call themselves *ekklēsia*, he only refers to the Philippian *ekklēsia* in comparison with other assemblies with whom he had contact. The central term in this clause for understanding Paul's relationship with the Philippian community is *ekoinōnēsen*, another *koinōnia* cognate. The dominant image of this phrasing is a business partnership for which Paul has kept proper track of "receipts and expenditures," not a democratic civic assembly. For more comparison between *ekklēsia* and *koinōnia*, see Jennifer Quigley, *Divine Accounting: Theo-Economics in Early Christianity* (New Haven: Yale University Press, 2021), 40–44.

16. Julien M. Ogereau, *Paul's Koinōnia with the Philippians: A Socio-Historical Investigation of a Pauline Economic Partnership* (Tübingen: Mohr Siebeck, 2014), 216.

17. Joseph Marchal, "Slaves as Wo/men and Unmen: Reflecting upon Euodia, Syntyche, and Epaphroditus in Philippi," in *The People Beside Paul: The Philippian Assembly and History from Below*, ed. Joseph Marchal (Atlanta: Society of Biblical Literature Press, 2015), 141–76.

18. These ideas are developed further in Quigley, *Divine Accounting*, 16–33.

19. David Wilhite coined this term in a chapter on Tertullian's interpretation of Paul in *Ad uxorem*; Wilhite, "Tertullian on Widows: A North African Appropriation of Pauline Household Economics," in *Engaging Economics: New Testament Scenarios and Early Christian Reception*, ed. Bruce W. Longenecker and Kelly D. Liebengood (Grand Rapids, Mich.: Eerdmans, 2009), 222–24. Wilhite calls for a resistance against "modernist dichotomies . . . between religion and economics," to "read Tertullian as very much concerned with both economic issues and religious ones—the line between the two often being imperceptible" (224). While I agree with Wilhite's move to resist modernist dichotomies, his analysis is largely limited to human transactions, in this case; the divine is largely overlooked.

20. Devin Singh describes a "theopolitical economy" to "denote the ongoing interaction between theological discourse and the political and economic spheres." He uses this term as both a historical and conceptual marker; while the conceptual is primary in his analysis, he notes that these enmeshings have "concrete manifestations and [are] perhaps more observable in the ancient world"; Singh, *Divine Currency: The Theological Power of Money in the West* (Stanford, Calif.: Stanford University Press, 2018), 12. While Singh's project begins later than my analysis, his point is correct earlier, as well. Divine-human transactions do have concrete manifestations that are more ubiquitous and observable in a variety of textual and material evidence.

21. My interest and approach to this topic have been influenced by the work of feminist materialists such as Jane Bennett, who have helped us to understand the interconnectedness of people and the objects around them, to get at the "vital materiality" of nonhuman things such as power grids and garbage dumps—and, I would add, economic forces such as "the market." "Agentic capacity is now seen as differentially distributed across a wider range of ontological types"; Jane Bennett, *Vibrant Matter: A Political Ecology of Things* (Durham, N.C.: Duke University Press, 2010), 9, 21. See also Karen Barad, who argues that "matter and meaning are inseparable"; Barad, *Meeting the Universe Halfway: Quantum Physics and the Entanglement of Matter* (Durham, N.C.: Duke University Press, 2007), 3.

22. Sven Gunther, "Taxation in the Greco-Roman World: The Roman Principate," *Oxford Handbook Online*, https://www.oxfordhandbooks.com/view/10.1093/oxfordhb/9780199935390.001.0001/oxfordhb-9780199935390-e-38#oxfordhb-9780199935390-e-38-bibItem-9. For more on the differences between *tributa capitis* (poll tax) and *tributa solis* (land tax), see P. A. Brunt's "The Revenues of Rome," *JRS* 71 (1981): 161–72.

23. Shaye J. D. Cohen, Review of Marius Heemstra, *The Fiscus Judaicus and the Parting of the Ways*, https://www.biblicalarchaeology.org/reviews/the-fiscus-judaicus-and-the-parting-of-the-ways/.

24. For more on this model and its afterlives in early Christian systems of almsgiving, see Singh, *Divine Currency*, 64–65.

25. Gregory S. Aldrete, *Daily Life in the Roman City: Rome, Pompeii and Ostia* (Norman: University of Oklahoma Press, 2009) 197.

26. For more on the logistics and processes of grain shipments to Rome, see Paul Erdkamp, "The Food Supply of the Capital," in *The Cambridge Companion to Ancient Rome*, ed. Paul Erdkamp (Cambridge: Cambridge University Press, 2013), 262–64, and G. E. Rickman, "The Grain Trade under the Roman Empire," *Memoirs of the American Academy in Rome* 36 (1980): 261–75.

27. For more on poverty and food insecurity in the Philippian assemblies, see Richard Ascough, *Paul's Macedonian Associations: The Social Context of Philippians and 1 Thessalonians* (Tübingen: Mohr Siebeck, 2003), 118, and Peter Oakes, "The Economic Situation of the Philippian Christians," in Marchal, *People Beside Paul*, 76–79. For more on abundance and lack in Philippians 4, see Nasrallah, *Archaeology and the Letters of Paul*, 122–28.

28. Oakes, "Economic Situation," 161.

29. Javier Zamora, "Citizenship," https://lithub.com/citizenship-by-javier-zamora/.

30. For an example of this kind of work, see Marchal, *People Beside Paul*.

31. Paula Fredriksen, *Paul: The Pagans' Apostle* (New Haven: Yale University Press, 2017), xii.

Ambiguous, Amorous, Agonistic, Not Able: An Alternative to Adamant, Apathetic, Antagonistic, Able Society

EUNCHUL JUNG

No century in human history has been free of hatred and violence. As civilization has become more complicated and expanded, antagonism—with its accompanying destruction and exploitation—among neighboring and competing groups of people increased in intensity and frequency until, finally, the entire globe is under its sway. If a peaceful and open society is said to constitute the ideal form of democracy, tribalism that results in extreme polarity and political antagonism is diametrically opposed to it. So why has tribalism been endemic to human civilization? It has been so endemic that tribalism deserves to be declared a global pandemic. This question about tribalism's ubiquity is one we must ask to diagnose our society, struggling to be democratic only to become more and more undemocratic—that is, liable to political antagonism.

The present chapter psychoanalytically investigates the inner workings of tribalism—that is, how individuals' negative affects are employed to fuel tribalism, which consumes noble virtues upon which humanity has (claimed to have) built their civilization. The chapter then explores the prospects for an alternative, more democratic public that is so desperately needed in this divided and hostile world.

First, to show the ways in which individuals' existential anxiety gives rise to a tribalist political movement, Robert Corrington's useful distinction and comparison between "a natural community" and "a community of interpreters" is discussed. This distinction first appeared in *The Community of Interpreters*[1] and was later elaborated in his most recent book, *Nature and Nothingness*.[2] Corrington's critical analysis of tribalism and suggestion of a hermeneutic community will then be complemented and expanded by Catherine Keller's introduction of a new public of amorous agonism presented in *Political Theology of the Earth: Our Planetary Emergency and the Struggle for a New Public*.[3]

I subsequently move on to discuss counterintuitive ethics from two different but closely related strands: systems of ethics that are believed to free us from the spell of tribalism and help build a more democratic society. On the one hand, Byeong Chul Han in *The Agony of Eros* suggests a phenomenology of Eros—that is, an ethics of *being able not to be able*, to liberate us from our current meritocratic and exploitative society driven by the Logos principle.[4] On the other hand, Jack Halberstam in *The Queer Art of Failure* advocates and promotes his queer ethics of failure to encourage us to detach ourselves from the logic of capitalism, which equates success with maximization of profits by failing well and failing often.[5] The strong affiliation between tribalism and capitalism is clear. As we shall discuss later, if we can call capitalism a brand-new and most efficient engine of tribalism, the alternative ethics of inability that Han and Halberstam are suggesting could be at the same time an attempt to stop fueling the engine and a new heart for the alternative public that Corrington and Keller are proposing.

A NATURAL COMMUNITY AND POLITICAL ANTAGONISM

In *Nature and Nothingness*, Corrington clearly shows how individuals, when faced with identity crises, reshape their identities to protect them. Existential predicaments "shake and transform identity in all its forms . . . and threaten to undo the self completely."[6] In times of trouble, people may initially curl up to wait until the difficulties pass like a hedgehog. However, if these troubles continue, they may begin to experience "overwhelming anxiety" and "the constant fear of fragmentation and the loss of a stable personal contour,"[7] for which the hedgehog strategy is insufficient. In this case, they will develop a deep sense of insecurity, and anxiety and its twin sons—fear and phobia—will prevail. There arises a desperate demand for the release from their anguish, which then may cause simultaneous positive and negative transferences.

On the one hand, the positive transference occurs when one hands over one's autonomy to a powerful political leader (*Führer*, the father figure), projects the narcissistic part of oneself onto the group one belongs to, and identifies oneself with the group. Through this transference arises a strong sense of belonging among people, often of a defensive and compulsive nature, and such a sense of belonging develops into tribalism. The size of the group in which individuals can have a strong but defensive sense of belonging ranges from a small familial clan through an ethnic group, a nation-state, and even up to humanity, and it can be arbitrarily based on anything—bloodline, region, skin color, language, religion, etc. Such a strong group consciousness allays, though temporarily, the fear and despair of individuals' own fragmented and anxious situations and stabilizes their egos.

On the other hand, one may "unleash the negative transference onto the target group, e.g., Jews, people of color, or women."[8] In this case, one turns self-loathing toward the Other, to "demonize and castigate the Other as a sinful and hell-bound being."[9] One can obtain a new, stronger identity, required to overcome the anxious situation through these negative transferences. One of the most efficient and thus widespread forms of negative transference is scapegoating, which "reestablishes the inner health and vitality of the group."[10] It seems to be inevitable that a group of people will seek out the seemingly dangerous others and through this form a solid *we*-identity, forming the "us vs. them" mentality. Tribalism only survives by continually creating the category of the Other and annihilating its members.

Corrington cites these positive and negative transferences to explain the formation of group consciousness and the violent expression of it, which together form the foundation and/or manifestation of tribalism from a psychoanalytical perspective. The same phenomena addressed by Corrington were also investigated by René Girard in his seminal literary critique, for which the mimetic theory of desire and the scapegoat mechanism are the key components. He elucidates the ways in which a society is drowned in negative and destructive affects caused by the dynamics of desire among individuals.

Girard impressively reveals how anxiety is swiftly disseminated among individuals by copy-pasting others' desires through his unique analysis of fiction novels in *Deceit, Desire, and the Novel: Self and Other in Literary Structure*.[11] It is no wonder that anxiety, along with its two immediate results—fear and contempt—have disintegrating effects. Rampant anxiety can hold an entire group in its sway, and relationships among individuals are largely characterized by transference, countertransference, recognition struggle, projection, imitation, meaningless competition, and reckless efforts. When a group has reached the apex of anxiety and conflict, it desperately looks for an outlet for anxiety, and certain types of people are frequently pointed out and name-called. The negative feelings thus far accumulated in the group need to be imputed to these unlucky few. In *Violence and the Sacred*, Girard traces the origin and vestiges of the scapegoating mechanism through a critical interpretation of religion, myth, psychoanalysis, and cultural rites such as marriage.[12] The process through which a group of people is completely devoured by negative affect and consequently in need of a scapegoat appears to be a part of our most typical and easiest— and probably the most efficient, at least in the short term—reaction to existential anxiety. In this regard, Corrington names such a defensive and compulsive group a natural/inert community.

In such a community, what Wilhelm Reich called "emotional plague" takes over in times of identity crisis, and this plague can "roar through a community (tribe) through its collective unconscious at astonishing speed."[13] One can secure "an ersatz counteridentity that generates emotional armoring and sets the self on the path to possible fundamentalism, which can be secular as well as religious."[14] Arguably, this typical and violent response of a natural community often goes together with a literal reading of sacred texts,[15] and Corrington warns that such a textual literalism is "a form of pathology that can eat away at the structures of democracy"[16] and gives rise to "patriarchy that deprives the self of any freedom and creativity."[17]

> Cultural barbarism must use physical violence against its perceived foes. Remember that textual literalism, the Führer principle, top-down theism, unconscious negative projections of one's own rejected half, and the emotional plague whipped up through propaganda (never a true art form), collectively make cultural barbarism into an ugly reality.[18]

The natural community, when drowned in the emotional plague, is likely to fall prey to polarity thinking, believed to generate dualistic clarity in times of uncertainty and anxiety. False dichotomies rule for the sake of quick and easy comfort for the ego.

In *Political Theology: Four Chapters on the Concept of Sovereignty*, Carl Schmitt famously defined the political as an antagonism between friend and enemy, and such political antagonism accurately reflects our natural/inert, though undesirable, attitude toward the Other, which, as discussed earlier, ends up creating tribalism.[19] From the perspective of evolutionary psychology, a group of individuals could increase its chances of survival by choosing either fight or flight. When it seems to be impossible to escape the apparently threatening Other, they would feel obliged to confront and antagonize the Other in their all-or-nothing competition for survival and hegemony—they could, of course, decide to surrender to the Other. Corrington points out that to carry on the war efficiently, they are often desperate for a powerful patriarch, the Father-*Führer*, the sovereign exception. Corrington notes that individuals tend to submit themselves to the *Führer* in order to "enforce and sustain a loyal and theistic form of obedience" to him, and "autonomy collapses under the weight of *heteronomy*, which is the imposition of alien law from above."[20] In the natural community, a homogeneous society is often ensued by heteronomous mass, which is ruled by dualistic thinking and does not allow room for open inquiry, critical reflection, and non-exclusive cooperation, all of which are believed to be integral to a better democracy.

A COMMUNITY OF INTERPRETERS AND AMOROUS AGONISM

As opposed to a natural community of political antagonism, there is an alternative community that responds to devouring existential anxiety in different ways. Corrington calls such a community "a community of interpreters" in the sense that the members of such a community tend to be *hermeneutically* open to and mindful of their changing and sometimes challenging environment and others. While one needs a new, stronger, more stable ersatz self in a natural community, one may learn to be content with an ambiguous, unstable, diffused self in the community of interpreters.

In such a mindful and self-reflective community, people do not need to seek the dualistic clarity often accompanied by mindless either/or thinking. For them, a creative encounter with identity crises is seen as "an invitation to . . . further growth and healthy . . . Selving process" that does not take on "the ersatz identity under the *Führer* principle,"[21] and therefore, the community of interpreters becomes "patriarchy-free."[22] This *unnatural* community keeps trying to embrace anxious situations and practice mindfulness required to "peel away scapegoating, the emotional plague, and the ubiquitous negative transference."[23] Accordingly, while natural communities often "lend themselves to barbarism," the community of interpreters is vigilantly on the lookout for it because the community members are aware that barbarism suffocates freedom and jeopardizes peace.[24]

Corrington's idea of the community of interpreters largely draws upon John Dewey's pragmatism, which understands democracy as experimentation that continually seeks betterment rather than perfection. In *A Common Faith*, Dewey argues that, for a better democracy to come, we should be attentive to the possible and the unknown rather than the actual and the known.[25] In *The Task of Utopia: A Pragmatist and Feminist Perspective*, Erin McKenna notes that a Deweyian understanding of democracy is "an ongoing task rather than a resting place"[26] and that progress, for Dewey, is to be found in "the ongoing activity of people seeking meaning in a changing world."[27]

The natural/inert community shows the tendency to remain stable and unchanging and therefore foreclose the possibilities of change once its members believe they have reached completion. On the contrary, the community of interpreters is fluid and wants to "keep the possibility of change alive"[28] and, unlike the natural community, subjects itself to "ambiguity," "imperfection," "difficulties," "inconsistencies," and "faults."[29] Corrington's community of interpreters is consistent with such a view of a democratic society as a "fluid association that encourages people to become Dewey's unified individual" and does not seek "a specific arrangement of society but a critical, flexible, and open-minded citizenry."[30]

Keller's public of amorous agonism adds an emotional aspect to Corrington's hermeneutic community. While the latter focuses on the continual psychological growth of individuals and their community through an ongoing, critical self-reflection, the former is formed through affectional solidarity and proactively engages social issues to respond to urgent demands of the suffering. Keller distinguishes "amorous agonism" from "political antagonism" in search of a new public in times of planetary emergency. Her new public of amorous agonism is new because it is an *unnatural*, hardly found type of community in evolutionary history. In other words, this public is formed out of sympathy rather than antipathy. It is motivated not by political ideologies but by empathizing with the vulnerable through an unexpected encounter with a series of tragic phenomena. As Rebecca Solnit writes, "When all the ordinary divides and patterns are shattered, people step up—not all, but the great preponderance—to become their brothers' keepers. And that purposefulness and connectedness bring joy even amidst death, chaos, fear, and loss."[31]

On March 27, 2020, Pope Francis delivered the extraordinary *Urbi et Orbi* [To the City (Rome) and the World], which was a deeply consoling and hopeful message, and prayed for an end to the coronavirus. He spoke to salvation through contingency and emergency, as expected, but his message went beyond providing spiritual comfort and hope. He had chosen the story of Jesus calming the stormy sea of Galilee in Mark 4 and took advantage of it to critique the death-dealing ways of life in the contemporary capitalist world. He stated that the uncontrollable storm "uncovers those false and superfluous certainties around which we have constructed our daily schedules, our projects, our habits and priorities" and exposes the superficiality of all our initiatives, thus removing the camouflage of our morbid, narcissistic ego investing in the capitalist and tribalistic culture. Pope Francis went on to contend that the global pandemic would end up bestowing salvation called solidarity and a sense of belonging to us stuck in "all our prepackaged ideas and forgetfulness of what nourishes our people's souls": "That (blessed) common belonging, of which we cannot be deprived: our belonging as brothers and sisters."[32]

Halberstam also believes that our inevitable epic failure—individual or collective, while being "the source of misery and humiliation," may lead to "a kind of ecstatic exposure of the contradictions of a society obsessed with meaningless competition . . . [and] the precarious models of success by which American families live and die."[33] While loss and failure first come to us as calamity, Halberstam also believes they may generate in us solidarity, cooperation, and compassion. The frightfulness of our violence-laden world may convert our individual sob and solitary groan into communal lament and unexpectedly summon us to provide sustaining, life-giving power for one another.

Keller maintains that such a new public resists the resentment—among races, nations, religions, regions, sexes, genders, classes, and generations—that "fuels antagonism and its spirit of retaliation."[34] Emerging out of emergency, the two different types of public—that of political antagonism and that of amorous agonism—are diametrically opposed in their response to the exigency. The public of amorous agonism is not a lasting community sharing common ideological agendas and articles of political faith, but it is rather a society of individuals who are entangled with each other emotionally—not always cohesive yet sporadically made cohesive by an emergent situation. Certain situations that "catalyze a mass mindfulness"[35] make a loosely combined ordinary public a cohesive and transformative political force. Its primary concern, at least initially, is not politics itself; rather, it is driven into the public space by compassion and caring for suffering bodies—human and nonhuman—and the devastated earth, our shared home.

It is important to note here that Keller's public of amorous agonism does not lack anger and forceful actions, though, if possible, these actions are nonviolent. Its agony can turn into agonistic resistance and produce transformative force: "The amorous agonism does not cease to blow or break into renewed political potency."[36]

In the community of interpreters, people embrace an identity crisis as an invitation to further psychic maturation and hope to contribute to the construction of a more inclusive, more democratic, and more pluralistic public by taking the risk of losing themselves. Keller's public of amorous agonism is formed in exigency and calls forth loving-kindness for and cooperation with others. Radical relationality underlies her process-style ecological ethics: we are all entangled in the karmic process. Both Corrington's and Keller's communities are immune to tribalism and political antagonism by virtue of their orientation toward an ever-enlarging self-identity. In other words, they keep moving beyond the narrow and exclusive ego—whether individual or collective—boundaries to include all fellow creatures so that none remains as *the Other*.

Both the community of interpreters and the public of amorous agonism are *apolitical* in the Schmittian sense of the political. Since they are apolitical, they may be politically incapable because their political skills are not as sophisticated as those of political communities in fulfilling a set of goals. Theoretically sought after as they may be, those politically incapable communities will perhaps fail in wild political reality. However, they should be willing to take the risk of being incapable. They should believe that only resistance through failure can bring about liberation from tenacious tribalism and political antagonism.

Now we shift our discussion from tribalism to capitalism to see how loss, failure, and incapability in capitalist society bring forth—or at least point to—a new prospect for alternative ethics beyond the snare of antagonistic, triumphalist tribalism. When our socio-politico-economic world is governed by a capitalism that propels tribalism, critique and resistance against capitalism will impair the efficiency of the engine of tribalism enough to prevent its acceleration.

A CRITIQUE OF CAPITALISM

Capitalism has unbridled tribalism and given it a firm footing. It goes without saying that tribalism never shrank under any ideology or economic system. But capitalism justifies unbound and unchecked competition among individuals and groups by equating success with virtue and failure with vice. Such capitalist work ethics allow us to strive to win against others by any means necessary without the slightest sense of guilt. Infinite competition among groups and individuals always divides winners and losers. Nation-states and individuals alike are assigned certain ranks by various standards and measures. Group resources are mobilized to antagonize competing groups. The group becomes increasingly homogeneous because the individuals want their group to be united, unified, and pure. The Others identified as threatening to the purity of this uniform group are persecuted or expelled. The principle of limitless competition is applied to every phase of individuals' lives. Free competition, success, reward, growth, achievement, and positivity work as solemn and self-evident axioms for work and love.

Both Han and Halberstam look for "ways of being and knowing that stand outside of conventional understandings of success . . . in a heteronormative, capitalist society [that] equates too easily to specific forms of reproductive maturity combined with wealth accumulation."[37] Han defines today's neoliberal capitalist society as an "achievement society." Since the dawn of civilization, human societies have advanced in ways that maximize productivity and inevitably accompany violent exploitation of the Other. Compared to pre-industrialized eras, today's capitalist society has attained unprecedented productivity. According to Han, it has skillfully blended the myth of voluntary individuals and motivation for achievement. Therefore, compulsive self-propelled exploitation of oneself has replaced involuntary labor and has become an integral part of today's work ethic.

> Achievement society is wholly dominated by the modal verb *can*—in contrast to disciplinary society, which issues prohibitions and deploys *should*. After a certain point of productivity, *should* reaches a limit.

To increase productivity, it is replaced by *can*. The call for motivation, initiative, and projects exploits more effectively than whips and commands.³⁸

Han calls those who have internalized, consciously or unconsciously, the achievement society's norms an "entrepreneur of the self" or the "achievement-subject." Although people believe themselves to be voluntary, they exploit themselves to accomplish the maximum productivity of society. They are self-galvanized and are unaware of the fact that they are not in service of themselves but instead the achievement-society.

> The exploiter is the exploited. The achievement-subject is perpetrator and victim in one. Auto-exploitation proves much more efficient than allo-exploitation because it is accompanied by a feeling of liberty. This makes possible exploitation without domination. . . . The paradoxical imperative, *Be free* . . . plunges the achievement-subject into depression and exhaustion. . . . The neoliberal regime conceals its compulsive structure behind the seeming freedom of the single individual, who no longer understands him- or herself as a subjugated subject ("subject to"), but as a project in the process of realizing itself.³⁹

Halberstam believes that capitalist societies wholly oriented to achievement are backed by positive thinking characteristic of North Americans. Positive thinking is for Halberstam "a North American affliction" and "a mass delusion that emerges out of a combination of American exceptionalism and a desire to believe that success happens to good people and failure is just a consequence of a bad attitude rather than structural conditions."⁴⁰ North Americans' overly positive face and language can make non-North Americans' ordinary and calm life look almost depressed.

Despite the conscious effort of the achievement-subject to pursue positive facial expressions and mindset, their unconscious and their bodies are irresistibly sensitive to the fear of loss and failure. As Halberstam indicates, loss and failure are inevitable to any system,⁴¹ and we can never be forgetful of them no matter how much we are trying to avoid them. And as Han points out, "Capitalism *only* works with debt and *default*."⁴²

While many thinkers such as Walter Benjamin view capitalism as religious, Han posits that capitalism cannot be religious because it does not know the usage of forgiveness—that is, exemption from debt and default. There is no end or exit in the vicious, debt-creating society. Exhausted and anxious, the achievement-subject bears all the responsibilities for its debt. Restlessly bluffing, the achievement-subject is indeed fragile and laboriously and vainly convinced that it can avoid unavoidable failure.

Depression is an increasingly common illness, and the number of sleepless nights in highly industrialized and affluent cities has grown. While North American achievement-subjects and their imitators complain about a variety of mental sicknesses and suffer from neurotic and perverse symptoms, their problems are never accepted or solved but suspended. Anxiety, like a liquid fear, visits and revisits the achievement-subjects and haunts their lives. They are momentarily comforted by shopping, entertainment, and therapies and then go to work with positive faces and language again. Those who have dropped out of this race end up on the street hopelessly (and homelessly) in the hospital, jail, or concentration camps.

> That is the ruse: now whoever fails is at fault and personally bears the guilt. No one else can be made responsible for failure. Nor is there any possibility for pardon, relief, or atonement. . . . The impossibility of mitigation and atonement also account for the achievement-subject's depression. Together with burnout, depression represents an unredeemable failure of ability—that is, it amounts to *psychic insolvency*.[43]

To speak in relation to tribalism when the achievement-subjects are exposed to excessive stress and the competition and conflicts among them are escalated, persons unfailingly seek out the Other and submit themselves to a charismatic leader to overcome the Other. Individuals need to experience collective success to elevate their abject egos. The collective success and the denouncement of the victims provide the individuals with an identity as a winner.

FAILURE OR BEING ABLE NOT TO BE ABLE

Halberstam's literary criticism of cultural products and artworks shows "the dignity of failure"[44] and proposes "an alternative vision of life, love, and labor,"[45] standing in contrast to "the grim scenarios of success that depend upon 'trying and trying again.'"[46] Failure, not success, is celebrated here as "a way of refusing to acquiesce to dominant logics of power and discipline and as a form of critique."[47] Halberstam's animating work, as he himself notes, is "a catalogue of resistance through failure."[48] Halberstam encourages us to be "nonbelievers outside the cult of positive thinking . . . the failures and losers, the grouchy, irritable whiners who do not want to 'have a nice day' and who do not believe that getting cancer has made them better people" and who will be relieved of "the obligation to keep smiling through chemotherapy or bankruptcy."[49]

As capitalism and tribalism conspire to beautify success and constant growth, every activity in life, including sex, is regarded as a form of labor for productivity and reproductivity, and society uses all means at its disposal to raise its members to be effective and successful workers. While what Halberstam calls "heteronormative common sense" has become the norm, his "subordinate, queer, or counter-hegemonic modes of common sense" lead us to "the association of failure with nonconformity, anticapitalist practices, nonproductive lifestyles, negativity, and critique."[50] In other words, his counterintuitive queer ethics of failure serves as "a refusal of mastery, a critique of the intuitive connections within capitalism between success and profit, and as a counterhegemonic discourse of losing."[51] Indeed, his book is all about "failing well, failing often, and learning, in the words of Samuel Beckett, how to fail better,"[52] and he believes that only in losing can we imagine goals for life and love other than productivity and reproductivity.[53]

Failure, of course, causes negative affects, but it may also crack open the "dogged Protestant work ethics"[54] and offer "unexpected pleasures"[55] by offering different rewards than traditional masculine success in the achievement society: "This failure, hilarious in its execution, poignant in its meaning, and exhilarating in its aftermath, is so much better, so much more liberating than any success that could possibly be achieved in the context of a teen beauty contest."[56] In other words, failure may allow us to escape "the punishing norms that discipline behavior and manage human development with the goal of delivering us from unruly childhoods to orderly and predictable adulthoods," the goal set by the achievement society for maximum productivity.[57]

Failure preserves some of the wondrous anarchy of childhood and disturbs the supposedly clean boundaries between adults and children, winners and losers. And while failure certainly comes accompanied by a host of negative affects, such as disappointment, disillusionment, and despair, it also provides the opportunity to use these negative affects to poke holes in the toxic positivity of contemporary life.[58]

"So much more liberating" as it may be for us individually, what will happen *among* us when we fail? Will we no longer need to view others as competitors that we should grasp, overcome, control, and consume if we let go of our obsession with success and try to embrace failure instead of exerting a tremendous, Herculean effort to succeed? Instead of looking at each other, tense, striving against loss, will we not probably choose to collaborate? Halberstam anticipates that "under certain circumstances failing, losing, forgetting, unmaking, undoing, unbecoming, not knowing may in fact offer more creative, more cooperative, more surprising ways of being in the world."[59] If we *undiscipline* ourselves to reinvent love, knowing, and work so we see our failure—inevitable or intended—as an asset or a style in resisting that which is and imagining that which can be and ought to be, we will then possibly relinquish "the Darwinian motto of winners" and cleave to "a neo-anarchistic credo of ecstatic losers: No one gets left behind!"[60] We will be able to weaken the capitalist norms that sharply divide winners and losers and corrode the engine of tribalism so that our society gets a little closer to a better state of democracy that many of us are envisaging.

As opposed to the capitalist way of relating to one another, Halberstam's queer art of failure invites us to "lose ourselves in its avenues of charming ignorance and spectacular silliness"[61] to make "an *unknowing relation* to the other."[62] Han also prescribes a way of relating to the Other that is similar to Halberstam's queer ethics of failure. Han's phenomenology of Eros is a call for an ethics of *being able not to be able* (Nicht-Können-Können) in relation to the Other. That is, it subverts the protestant/capitalist work ethics and completely changes the ways in which we relate to the Other. This is not an innocent inability but something like learned ignorance, intentional retreat from our fixation on performance. This is a characteristic of Eros that is opposed to the culture hero Prometheus.

The Promethean achievement-subject and society will never be able to escape tribalism unless they withdraw from their fetish of achievement and stop consuming the Other—human or nonhuman. The Other is easily sacrificed for the sake of our tribe's prosperity. But the erotic subject incapacitates itself from consuming the Other so that the Other manifests itself on its own terms. Only to the incapable, erotic subject reveals the Other its unique countenance. Eros requires more than coexistence and cooperation. As Alain Badiou writes in his Preface to *The Agony of Eros*, "The minimum condition" for true Eros is "possessing sufficient courage to accept self-negation for the sake of discovering the Other," so that negativity of the Other unshackles the ego from the logic of the achievement society.[63]

> Eros is a relationship to the Other situated beyond achievement, performance, and ability. *Being able not to be able* . . . represents its negative counterpart. The negativity of otherness—that is, the atopia of the Other, which eludes all ability—is constitutive of erotic experience. . . . A successful relationship with the Other finds expression as a kind of *failure*. Only by way of *being able not to be able* does the Other appear.[64]

The erotic subject is the subject who renounces its addictive engagement with achievement and ability and who is thus *able not to be able* to possess the Other. Here, the Other disastrously invades the achievement-subject, and this is an experience of "a catastrophe for the ordinary balance of the subject"—that is, the loss of subjectivity. However, this apparently disastrous irruption of the Other turns out to be "the way to redemption" that transforms the achievement-subject into the erotic one, a redemption not *of* the ego but *from* the ego and its constructs.[65]

CONCLUSION

Corrington's community of interpreters deals with existential crises in constructive and flexible ways by being mindful of the ego's unconscious defensive mechanism instead of having recourse to it pathologically and voraciously. Keller's public of amorous agonism is formed and prompted not by political aspirations, animosity, and ideological schemes but by unconditional compassion for all creatures. Both these communities are antithetical to tribalism and political antagonism to which a natural/inert community is prone. Both Halberstam's queer ethics of failure and Han's Eros ethics of *being able not to be able* contradict capitalist work ethics and its way of relating to the Other, both of which are buttressed by the meritocratic doctrines of limitless competition and constant growth and serve as the propeller of tribalism.

However, will failure really enable us to love and cooperate freely and non-exclusively instead of resorting to compulsive antagonism in fear of uncertainty? Will we not employ tribalism and antagonism again to *achieve* our set ideals for a better democracy? Schmitt's definition of the political as an antagonism between friends and foes, though much criticized, accurately explains the past and present—and surely the future—political landscape in its *natural* setting. Even many progressive resistant movements seem to compromise on their methods for their ideals. Antagonistic identity politics and its ever-increasing, irreconcilable polarities have only made the situation worse and therefore become the disguised and most skillful perpetrators or unintended perpetuators of the patriarchal, phallic, tribalistic, belligerent, phobic, exploitative, destructive, neoliberal, capitalist civilization.

Antagonism and triumphalism seem to have become our political genes. Some cultural theorists, such as Herbert Marcuse, have pointed out that as long as they are driven by political antagonism and ideological agendas, even progressive movements could harm democracy and instead serve tribalism. Even democracy itself—in its characteristic partisanship, growing polarity, and winner-takes-it-all electoral system—appears to be "the most efficient system of domination."[66] Furthermore, for Marcuse, every revolution in human history has turned out to be "a betrayed revolution" in that it was an attempt to build a better system of domination:[67] "Liberty follows domination—and leads to the reaffirmation of domination."[68]

Halberstam openly rejects the antagonistic feminist resistance, refusing "triumphalist accounts of gay, lesbian, and transgender history that necessarily reinvest in robust notions of success and succession,"[69] which only perpetuate the patriarchal system in the guise of *masculine* femininity. Instead, he suggests a "shadow feminism" rather than a "more acceptable one,"[70] whose motto would be "Practice more failure."[71] Halberstam has noticed that any form of resistance that antagonizes and tries to win against the opponents will eventually reinforce triumphalism and meaningless competition. Even though successful resistance is too attractive to resist, we should refuse it. For, as Michel Foucault pointed out, it is a good time to look at ourselves to see if we resemble our opponents the most when we are preoccupied by what we are doing.

> How does one keep from being fascist, even (especially) when one believes oneself to be a revolutionary militant? How do we rid our

speech and our acts, our heart and our pleasures, of fascism? How do we ferret out the fascism that is ingrained in our behavior?[72]

We should always be attentive to whether the grammar of the system that we are resisting remains operative in ourselves. Otherwise, we may be resisting only to strengthen and eternalize the logic of the achievement society where antagonism and tribalism rage. Only the ruling classes alternate, and the dominant system persists. Our refusal of its grammar will be the most acute subversion of the system, just as Jesus's failure was reversed into the divine achievement, and his *being able not to be able* was the most powerful divulgence of and accusation against violence deeply seated in the human psyche and civilization.

Keller notes that the crucified Jesus is crossed over by two different types of politics: the sovereign antagonism and the amorous agonism. "The former nails above the suffering body of Jesus a sign, *The King of the Jews*, in sovereign ridicule of the messianic promise; the latter inspires a long history of struggles for the *basileia theou*, the kingdom of the least, the parody of power," and Jesus's crucifixion "endlessly recalls an agony that cannot be erased."[73] Therefore, Jesus's life, death, and Kingdom of God movement proclaimed the inception of weak, amorous agonism rather than the omnipotent sovereign exception, and his weak, amorous "messianism is the red secret of every revolutionary."[74] Will we be able to live up to Jesus's most radical teaching—"Love your enemy"—which we, Christians or not, have tried to avoid so much as we have admired?

What shall we then do in search of "the possibility of other forms of being, other forms of knowing, a world with different sites for justice and injustice?"[75] As frequently seen in many of what Halberstam calls *Pixarvolt* (Pixar + revolt) films, we should perhaps learn from the silly animal characters to fail well, fail often, and *be able to not to be able*.

> To live is to fail, to bungle, to disappoint, and ultimately to die; rather than searching for ways around death and disappointment, the queer art of failure involves the acceptance of the finite, the embrace of the absurd, the silly, and the hopelessly goofy. Rather than resisting endings and limits, let us instead revel in and cleave to all of our own inevitable fantastic failures.[76]

All the previously discussed alternative communities and ethics only dwell in "the murky waters of a counterintuitive, often impossibly dark and negative realm of critique and refusal."[77] Our dream world will not be another, more steadfast system realized by competing, winning, and achieving, and this is what McKenna would call "the end-state model of utopia," which is identical to "totalitarianism."[78] Rather, it will be a certain middle state that is allowed for us to live right now only when we surrender our ego constructs and let loose from our natural political inclination—antagonism and tribalism—and this is what McKenna would call "the process model of utopia." We should dream of an already existing future, not the one that is yet to come and somewhere else, and that we should thus try and try again to get there. We are not liberated by achieving another, more advanced, more progressive system in the future. We bring truly liberated life into the present by not belonging to any system—that is, by failing and by *being able not to be able*.

NOTES

1. Robert S. Corrington, *The Community of Interpreters*, 2nd ed. (Macon, Ga.: Mercer University Press, 1996).
2. Robert S. Corrington, *Nature and Nothingness: An Essay in Ordinal Phenomenology* (Lanham, Md.: Lexington Books, 2017).
3. Catherine Keller, *Political Theology of the Earth: Our Planetary Emergency and the Struggle for a New Public* (New York: Columbia University Press, 2018).
4. Byung-Chul Han, *The Agony of Eros* (Cambridge, Mass.: MIT Press, 2017).
5. Jack Halberstam, *The Queer Art of Failure*, Illustrated ed. (Durham, N.C.: Duke University Press, 2011).
6. Corrington, *Nature and Nothingness*, 96.
7. Corrington, *Nature and Nothingness*, 2–3.
8. Corrington, *Nature and Nothingness*, 3.
9. Corrington, *Nature and Nothingness*, 8.
10. Corrington, *Nature and Nothingness*, 10.
11. René Girard, *Deceit, Desire, and the Novel: Self and Other in Literary Structure* (Baltimore: Johns Hopkins University Press, 1976).
12. René Girard, *Violence and the Sacred*, trans. Patrick Gregory, 1st ed. (Baltimore: W. W. Norton, 1979). This work echoes the mood evoked by the raw and unresolved sound of Igor Stravinsky's ballet *The Rite of Spring* (1913)—especially the third episode of part 2, "Glorification of the Chosen Victim."
13. Corrington, *Nature and Nothingness*, 1.

14. Corrington, *Nature and Nothingness*, 113.
15. Corrington, *Nature and Nothingness*, 6.
16. Corrington, *Nature and Nothingness*, xiii.
17. Corrington, *Nature and Nothingness*, 113.
18. Corrington, *Nature and Nothingness*, 99.
19. Carl Schmitt, *Political Theology: Four Chapters on the Concept of Sovereignty* (Chicago: University of Chicago Press, 2010).
20. Corrington, *Nature and Nothingness*, 99; author's italics.
21. Corrington, *Nature and Nothingness*, 11.
22. Corrington, *Nature and Nothingness*, 19.
23. Corrington, *Nature and Nothingness*, 10.
24. Corrington, *Nature and Nothingness*, 69. It should be noted that the adjectives "natural" and "unnatural" used in this chapter are value-neutral, even though I portray a natural community as a pejorative and an unnatural community as a compliment in the context of confronting indelible violence in human nature. In addition, I do not mean by the "unnatural" something outside of nature but only things that seem to run counter to our more typical political inclination.
25. John Dewey, *A Common Faith*, 2nd ed. (New Haven: Yale University Press, 2013).
26. Erin McKenna, *The Task of Utopia: A Pragmatist and Feminist Perspective* (Lanham, Md.: Rowman & Littlefield, 2001), 3.
27. McKenna, *Task of Utopia*, 6.
28. McKenna, *Task of Utopia*, 6.
29. McKenna, *Task of Utopia*, 9.
30. McKenna, *Task of Utopia*, 135.
31. Rebecca Solnit, *A Paradise Built in Hell: The Extraordinary Communities That Arise in Disaster* (London: Penguin, 2010), 3.
32. Pope Francis, "Pope at Urbi et Orbi: Full Text of His Meditation—Vatican News," March 27, 2020, https://www.vaticannews.va/en/pope/news/2020-03/urbi-et-orbi-pope-coronavirus-prayer-blessing.html.
33. Halberstam, *Queer Art of Failure*, 5.
34. Keller, *Political Theology of the Earth*, 26–27.
35. Keller, *Political Theology of the Earth*, 91.
36. Keller, *Political Theology of the Earth*, 54.
37. Halberstam, *Queer Art of Failure*, 2.
38. Han, *Agony of Eros*, 9; author's italics.
39. Han, *Agony of Eros*, 9–10; author's italics.
40. Halberstam, *Queer Art of Failure*, 3.
41. Halberstam, *Queer Art of Failure*, 94.
42. Han, *Agony of Eros*, 11; author's italics.

43. Han, *Agony of Eros*, 10–11; author's italics.
44. Halberstam, *Queer Art of Failure*, 92.
45. Halberstam, *Queer Art of Failure*, 2.
46. Halberstam, *Queer Art of Failure*, 3.
47. Halberstam, *Queer Art of Failure*, 88.
48. Halberstam, *Queer Art of Failure*, 97.
49. Halberstam, *Queer Art of Failure*, 4.
50. Halberstam, *Queer Art of Failure*, 89.
51. Halberstam, *Queer Art of Failure*, 11.
52. Halberstam, *Queer Art of Failure*, 24.
53. Halberstam, *Queer Art of Failure*, 88.
54. Halberstam, *Queer Art of Failure*, 96.
55. Halberstam, *Queer Art of Failure*, 4.
56. Halberstam, *Queer Art of Failure*, 5.
57. Halberstam, *Queer Art of Failure*, 3.
58. Halberstam, *Queer Art of Failure*, 3.
59. Halberstam, *Queer Art of Failure*, 2–3.
60. Halberstam, *Queer Art of Failure*, 5.
61. Halberstam, *Queer Art of Failure*, 59.
62. Halberstam, *Queer Art of Failure*, 12; my italics.
63. Alain Badiou, Preface to Byung-Chul Han's *The Agony of Eros* (Cambridge, Mass.: MIT Press, 2017), xii–xiii.
64. Han, *Agony of Eros*, 11; author's italics.
65. Badiou, Preface to Byung-Chul Han's *Agony of Eros*, xiii.
66. Herbert Marcuse, *One-Dimensional Man: Studies in the Ideology of Advanced Industrial Society* (Boston: Beacon Press, 1991), 69. William Connolly would also agree with this; see his *The Fragility of Things* (Durham, N.C.: Duke University Press, 2013).
67. Marcuse, *One-Dimensional Man*, 91.
68. Marcuse, *One-Dimensional Man*, 65.
69. Halberstam, *Queer Art of Failure*, 23.
70. Halberstam, *Queer Art of Failure*, 4.
71. Indeed, the whole notion of failure as a practice was introduced to me by the legendary lesbian performance group LTTR. In 2004 they asked me to participate in two events, one in Los Angeles and one in New York, called "Practice More Failure," which brought together queer and feminist thinkers and performers to inhabit, act out, and circulate new meanings of failure. Chapter 3, "The Queer Art of Failure," began as my presentation for this

event, and I remain grateful to LTTR for shoving me down the dark path of failure and its follies.
72. Michel Foucault, Preface to Gilles Deleuze and Félix Guattari, *Anti-Oedipus: Capitalism and Schizophrenia* (Minneapolis: London and New York: Continuum, 2008), xv.
73. Keller, *Political Theology of the Earth*, 55; author's italics.
74. Ernst Bloch, *Atheism in Christianity: The Religion of the Exodus and the Kingdom* (Brooklyn: Verso, 2009), 240.
75. Halberstam, *Queer Art of Failure*, 52.
76. Halberstam, *Queer Art of Failure*, 186–87.
77. Halberstam, *Queer Art of Failure*, 2.
78. McKenna, *Task of Utopia*, 1.

What Does Evolutionary Biology Tell Us about Relationality as a Basis for Economics and Politics?

MARCIA PALLY

This chapter was developed in the context of the 2020 Transdisciplinary Theological Colloquium, whose theme was the relationship between religion and economics, democratic politics, and environment. I begin with the premise that we cannot build an economics or politics for human flourishing until we know what sort of creature humans are, what contributes to our thriving in our ecological setting. The neurobiologist Darcia Narvaez writes, "To approach eudaimonia or human flourishing, one must have a concept of human nature, a realization of what constitutes a normal baseline, and an understanding of where humans are."[1] In short, one needs to know the nature of humanity, its ontology, to create conditions for its *eudaimonia* or thriving. The Greeks thought similarly: one studies natural philosophy for the "religious" purpose of learning how the natural and human world works—Narvaez's baseline—to live with it in harmony and near eudaimonia. "The nature of a thing," Aristotle writes in Book I of *Politics*, "is its end." To understand the nature of a thing is to understand its end, what counts as its specific form of flourishing.

What is our "baseline" or ontology so that we may develop an economics and politics to suit? Both Christian and Jewish traditions propose that it is relational, and it seems that evolutionary biology and developmental psychology are catching up to the idea. I'll begin with a discussion of relationality, drawing on the concepts of covenant and Trinity, and continue with a look at recent biological research on human cooperativity in hopes of developing a "baseline" that could serve as a framework for an economics and politics.

AN ONTOLOGY OF RELATIONALITY

We might begin by noting that developing personal and societal practices grounded in ontology is the central biblical import. "There exists a law," Yoram Hazony writes, "whose force is of a universal nature, because it derives from the way the world itself was made, and therefore from the natures of the men and nations in this world."[2] This foundation, the "way the world itself was made," is, in Jewish and Christian traditions, relational.

Relationality starts with the notion that Being, the possibility for existence, results from the source of all that is. There could be nothing, but there is something. The source of all "something"—items, thoughts, language, laws of physics—is what some people call God. Ian Barbour writes of God as a "structuring cause" or "designer of a self-organizing process."[3] Franz Rosenzweig called it "the eventfulness of the limitless possibilities that will come to exist, the not-nothing that is the 'divine essence in all infinity' prior to there being a distinct something or a distinct nothing."[4] After the kabbalist concept *Ein Sof* and F. W. J. Schelling, this source is not so much what precedes effects as what is realized as it yields effects. Existence, said briefly, is God's self-expression.

On one hand, each particular is radically different from structuring cause—differences in materiality/immateriality, finitude/infinitude, composite features/unitary simplicity—yet on the other, each particular partakes of it to exist at all. We are grounded by the source of existence in order to be, and that source grounds all particulars, essences, and features, be they past, present, or future. Yet, as Thomas Aquinas notes, we do not partake of the transcendent source identically or proportionally but rather analogically, as an analogy expresses its referent, with different features but an undergirding of-a-kindness. More specifically to Aristotle, analogical terms refer to *similarity of a feature that is present in both parties and accidental* to at least one of them. We are radically different from God yet with underlying of-a-kindness. The *b'tselem Elohim/imago* expresses this well: persons are radically different from incorporeal, imageless God; there are no features or divine physiognomy for humanity to partake of. Yet we partake analogically of the divine imageless "image."

Radical difference from the transcendent yet unavoidable partaking/relation is the way anything comes to be. *The structure of existence is difference-amid-relation.* Aquinas writes, "God himself is properly the cause of universal being which is innermost in all things [beings] . . . in all things

God works intimately."[5] All existing things share the property of radical distinction from the transcendent amid foundational partaking. "The One," in Catherine Keller's words, "is to multiplicity as white light is to the spectrum of a rainbow."[6]

As difference or distinction-amid-relation is the structure of existing, not only are persons distinct from God yet in intimate relation, but we are also distinct from each other yet in necessary relation. Aquinas understands it this way: we partake of God to exist, but we are not only distinct from God but from each other. Moreover, each one of us is a composite of various, distinct features or essences. Yet God, of whom we partake, is a simple unity and not various, neither a group of persons nor an aggregate of different essences. How is it that we, composites as individuals and distinct from each other, analogically partake of something simple, without components? It is possible, on Aquinas's account, because we are not only composites and distinct from each other but at the same time part of the unity that is God's self-expression.[7] "Thus," Mary Hirschfeld writes, "there must also be a unity to creation if creation is to give witness to the fact that God is one. God communicates this unity by ordering created beings to one another and all things to him. . . . These two features—the heterogeneity of created beings and their ordering *to one another* (and ultimately to God)—need to be respected."[8] This is our foundational, ontological relationality.

In the Jewish tradition, Emmanuel Levinas wrote, "My very uniqueness lies in the responsibility for the other man."[9] Echoing this, Martin Buber notes that "the individual is a fact of existence insofar as he steps into a living relation with other individuals."[10] In the Greek Orthodox tradition, John Zizioulas echoes, "The person cannot be conceived in itself as a static entity, but only as it *relates to* . . . [it is] in communion that this being is *itself* and thus *is at all*."[11] Or in Elisabeth Moltmann-Wendel's words, "Life begins as life together."[12] Kirk Wegter-McNelly, building on Wolfhart Pannenberg, summarizes: cosmos is "a place in which entangled independence-through-relationship is the fundamental characteristic of being."[13] Karl Rahner calls this "unity-in-difference,"[14] Catherine Keller and Laurel Schneider "entangled" or "non-separable" differences.[15]

Relationality as distinction-amid-relation is not a binary between distinction on one hand and relation on the other. It is rather reciprocal constitution: each becomes the singular, unique person she is through layers and networks of relations. It is the networks of interactions that constitute us. As that is our grammar or baseline, human flourishing entails that we see and *see to* the relations that are how we become who we are. At/tending to this baseline is the precondition for living in harmony in the world and nearing eudaimonia. The consequences of flouting it include cognitive and emotional impairment in children.[16] Adults who become isolated suffer from increased risk of suicide, mortality,[17] and morbidity, including depression and other emotional disorders.[18] The distresses resulting from the COVID isolation are a recent testament to humanity's relational constitution.

We become our (distinct) selves through relations with those nearby and through relations that extend out in our paths of global connectedness, as our educational and economic opportunities, nutrition and health care, and the tensions we and our relations are under are formed by those who are not necessarily geographically proximate. Contra social contract theory, there is no a priori individual who later enters a social contract because (singular) persons don't occur other than through their relations. Contra Kant, there is no autonomous lawgiver who individually reasons her way to universal precepts as reason develops through engagement with the ideas, practice, and methodologies of thought of other persons and cultures (as Polanyi, Kuhn, and others explain). Finally, relationality as distinction-amid-relation means neither homogeneity of persons nor of cultures but rather reciprocal commitment and responsibility among those who are different.

RELATIONALITY, TRINITY

The Trinity is a wonderful teacher of this idea. Each trinitarian person is distinct, each with "its own particular distinguishing notes," as Gregory of Nyssa wrote.[19] Yet each is who "he" is through relation to other trinitarian persons. Edith Stein, the German Jewish philosopher who became a Carmelite sister, notes that for the persons of the Trinity, "I am" is identical with "I am one with you" and with "we are."[20] The notion of *perichoresis*, as the Cappadocian fathers developed it, imagines the three trinitarian persons loosely "in a dance around," where the identity of each emerges from relation to the others. It is these relations that constitute the whole of the Godhead. "Person," John Milbank notes, is a relational term. "Yet this does not entirely collapse the persons into the relations, because 'person' is here rather the point of equipoise between relation and substance . . . (*ST* [*Summa Theologica*] I, q. 29, a. 4 resp.)."[21]

As it is each trinitarian person, distinct yet in relation, that constitutes the Godhead, without both distinction and relation, each is not a person *of the Trinity* because, as an isolated person, there is no Trinity—no unity—to be part of. In the Trinity, however, donative relationality transforms each from isolation into a person of divine communion. "The deity of this God," Wegter-McNelly writes, again building on Pannenberg, "resides not in the persons as distinct from one another but within and among the persons as they are related to one another, i.e., *in the relationality that constitutes* them and binds them."[22]

As God's self-expression in creation allows for all existence, we may say that the distinct trinitarian persons in mutual constitution allow for our human existence. We partake analogically of the "image" of the trinitarian community. Humanity, analogically partaking of the triune God, partakes of distinct-persons-in-relation.[23] On one hand, the distinction-amid-relation nature of God in himself (Immanent Trinity) informs how humanity understands God (God in relation to us, the Economic Trinity). God makes his communal (trinitarian) self-known to us in scripture and revelation. These are communicative, relational acts. On the other, the Immanent Trinity also renders each human being, in God's image, distinct-amid-relation (a creational and ontological act). As Aquinas held, the nature of the trinitarian God illuminates the human condition.[24]

In Pannenberg's elaboration, not only is the Trinity the ground for human relationality (a downstream flow, so to speak, from transcendent to humanity) but human relationality is inherent in the immanent Godhead (an upstream flow). After all, two of the three trinitarian persons, Son and Spirit, are who they are only in engagement with us. This makes our capacity for relation—with Son, Spirit, and each other—part of what it means for them to be Son and Spirit.[25] Indeed, part of what it means to be Trinity, for without the Son and Spirit, there is no Triune God. In a related reading of the triune *imago*, Jürgen Moltmann writes that the entire human community, not individual persons, is in the image of the communal God. It is not each person who is in God's image but rather persons together. As God is the unity of multiplicities, it is the union of multiple persons that is in his image.[26]

As each trinitarian person *gives* identity to the others, donation of a trinitarian type is without loss. Indeed, it is with repletion of identity. Thus, each human person, in the image of this donative God, is also more herself in the act of giving—an idea with not insignificant consequence for economics and politics. Aaron Riches and Daniel Bell are right to follow Anselm in rejecting the binary between caring for oneself *or* for another. Following trinitarian logic, Bell proposes that one may "overcome" the "modern illusion of the isolated, alienated self (or postmodern dissolute self)" for life "lived as donation . . . life as participation in the dance of charity that is the Trinity."[27]

RELATIONALITY, COVENANT

My second illustration of relationality is the concept of covenant, a bond between distinct parties where each gives for the flourishing of the other. Covenant, Jean Lee writes, is the "promise with one or more counterparty under common pursuit of shared values for long-term cooperation and well-being of the community." It is the promise, shared values, and *telos* of long-term communal well-being that distinguish covenant from other human transactions. Importantly, unlike contract, which protects interests, covenant protects relationship. Or, as Lee continues, "Contracts form the basis of the market while covenants form the basis of community."[28] In the Judaic tradition, the source of human covenantality is threefold. Most basically, we exist in the distinction-amid-relation grammar of existence; second, we are in the image of a covenant-making God (this is our nature); and third, we are in covenantal relation with God (this is the kind of relating we do, our activity). Stephen Geller writes, the Hebrew Bible God is not so much a concept, an "ism," as a relation.[29]

It's worth noting that while covenant creates community, it does not subsume the person, nor is the individual sacrifice-able for its sake (the point of the *Akedah*, the binding of Isaac narrative). In Lenn Goodman's words, "The covenant itself . . . rests on (and thus cannot create) the freedom of the covenantors."[30]

Covenants of reciprocal commitment among equals are easily imagined, as are covenants with asymmetric terms between unequal parties. The innovations of the Hebrew Bible are two: (1) Covenants of *mutual* commitment are forged between unequals, between the divine and human and among persons of different status; and (2) Covenant is not between lord and a vassal, who represents the people in his domain (as ancient suzerainty treaties were) but between God and each person directly.

The consequences of this intimate God-person bond include removing values, morality, and practices from the aegis of human monarchs and understanding them as grounded in the transcendent, whose values one cannot tweak to suit reigning political powers. Indeed, ordinary persons know the moral law and may judge the king's adherence to it. The Hebraic covenant, Robert Bellah writes, is "a charter for a new kind of people, a people under God, not under a king . . . a people ruled by divine law, not the arbitrary rule of the state, and of a people composed of responsible individuals."[31]

In this God-person covenantal reciprocity, stipulative features might arise (as parents stipulate that a child clean her room), but covenant is not stipulative in motive or *telos* (one doesn't have children so that they clean their rooms).

One aspect of covenant between God and humanity is its inauguration and maintenance by gift, often of an item of little economic value. As Marcel Mauss and Lewis Hyde have observed,[32] the spirit of the donor is given to the donee in the performative act of gift-giving. Donation of spirit makes the bond of trust, loyalty, and acknowledgment of a common future. This gift is neither contract nor quid pro quo. It does not imagine a transactional return, nor does it seek, however subtly, to coerce or manipulate. It is the mark of reciprocity for the sake of the other and shared horizons.

Covenantal donation of gift begins dyadically: God-Adam, God-Noah, God-patriarchs. The human partners in covenant are given the gifts of land and children, of survival in the world. In reciprocity, their children return a token of the land, of the harvest, to the Temple and God as a symbol of mutual commitment. Yet covenantal giving does not remain dyadic. Persons give to God also by giving in charity—in Hebrew, *hekdesh* (made holy). In this triangulation, one gives to God by giving to a third party, persons in need. These triune relations-of-giving are mutually constitutive: covenantal commitment to other persons constitutes covenant with God, and covenant with God sustains us in covenantal commitments to others. "Covenant is," Eric Mount explains, "a distinctively, though not exclusively, Hebraic metaphor and model that locates the relational self in a community of identity, promise, and obligation with God and neighbor."[33] The triangulated covenant is found in the Ten Commandments, the first three of which pertain to person and God, the rest, to life among persons. Amos and Proverbs go so far as to denounce the hypocrisy of performing rituals while abandoning the afflicted, as if one could maintain bond with God absent bond with the needy[34]—one of the most oft-repeated of biblical and rabbinic denunciations.

Covenantal giving thus extends from dyad to larger associations. Reciprocal giving becomes gift exchange network, as Mauss described it, where gift from God to person generates gift from person to person and on to the next person through the giving loop, thus sustaining it.[35] While gift-exchange networks exist within economic systems, they are the aspect of economic relations that marks mutual trust and shared fate, characterized by: (1) delay of return (immediate return feels like payment, not gift exchange); (2) nonidentical repetition (the returned item is never the same as the initial one); (3) recipient orientation, where the gift aims at benefiting not the giver but the recipient; and importantly, (4) asymmetrical reciprocity, wherein gift from A to B generates gift from B to C, etc. Person A receives a gift in return in the course of life and time after many gifts have traveled in multiple directions through the giving loop, thus maintaining it.[36]

Who is in the loop? Consistent with the idea that covenant/relationality is the structure of all existence, the biblical answer is: all the nations. The covenant to the patriarchs, thrice repeated, is "for the blessing of all the nations" (Gn 12:3, 26:4, 28:14). God covenants with non-Israelites as all persons, made in God's image, are capable of "moral correspondence" (*dmuth Elohim, similitudo*), of committing themselves to covenantal bonds and standards of relation. Such commitment undergirds the extensive biblical and rabbinic obligations to the enemy, stranger, as well as to the domestic poor.[37] The rabbinic *Mikhilta de-Shimon* (bar Yochai), commenting on Exodus 19:2, notes that the Torah was given not in any country but in the open desert to ensure access to all persons because its principles pertain to all. "The Torah speaks the language of human beings" (*b. Nedarim 3a, b. Berachot 31b*).

In the Christian Testament, the triangulation of covenantal commitment is seen in the famous passage in 1 John 4:20: "For whoever does not love their brother and sister, whom they have seen, cannot love God, whom they have not seen." Absent love of persons, there is no love of God. But, John continues, love of God enables love of others: "We love because he first loved us" (1 Jn 4:19). Love by God enables and sustains our love of other persons. Irenaeus put it concisely: "To love Him above all, and one's neighbor . . . do reveal one and the same God."[38] In Augustine's theology, the "relic" of God in each person gives her the capacity to love other persons. In sum, God makes to humanity a triune gift in covenant: the gift of relational existence, of being in God's (relational) image, and (in the Christian tradition) the gift of *a* relational being, himself in Jesus. In his image, we have the capacity (*similitudo, dmuth Elohim*) to respond in covenant to God and others.

The medieval period gives us one of the most soaring expressions of the triangulated covenant. The French-Jewish Bible commentator Rashi reads in Isaiah, "I cannot be God unless you are my witness," and Rashi glosses, "I am the God who will be whenever you bear witness to love and justice in the world." God can be God when persons are loving and just to each other. In the twentieth century, Levinas's work on responsibility to the "face" of the other expounds on this idea. "To follow the Most-High is also to know that nothing is greater than to approach one's neighbor."[39] This for Levinas is not a "figure of speech" but a description of God, "who approaches precisely through this relay to the neighbor—binding men among one another with obligation."[40] When Levinas writes that relationship with God "can be traced back to the love of one's neighbor,"[41] he does not start with God and derive responsibility to persons from covenant with the divine. Rather, the bond with God reaches back to commitment to neighbor. The philosopher Richard Kearney reprises: "This is a *deus capax* who in turn calls out to the *homo capax* of history in order to be made flesh, again and again—each moment we confront the face of the other, welcome the stranger." Echoing Levinas and Rashi, Kearney concludes that "welcoming the stranger" is the site of our bond with God: "A capacitating God who is capable of all things cannot actually be or become incarnate until we say yes."[42]

RELATIONALITY, EVOLUTIONARY BIOLOGY, DEVELOPMENTAL PSYCHOLOGY

What do the physical sciences tell us about relationality as a human "baseline"? In research on cognitive development and societal formation, evolutionary psychology and biology identify *H. sapiens* as a "hypercooperative" species.[43] Cooperative behaviors "are associated with a disadvantage or cost for the actor and a benefit for the recipient."[44] While evolutionary pressures yielded episodic aggression and opportunistic raiding where advantageous, cooperativity and egalitarianism (including communal property and childcare), along with robust fairness and sharing norms, were the *modus vivendi* of "modern" hunter-gatherers for 250,000 or so years, until roughly 8,000 BCE. Christopher Boehm describes the emergence of hunter-gatherer egalitarianism from our far less cooperative and more aggressive primate ancestors so that "over time, the apelike, fear-based, ancestral version of personal self-control would have been augmented, as there appeared some kind of a protoconscience that no other animal was likely to evolve."[45]

RELATIONALITY AND HUMAN COGNITIVE AND SOCIAL DEVELOPMENT

Human cooperativity was evolutionarily advantageous not only for survival reasons, such as equitable resource distribution that allowed for greater longevity and thus increased chances to reproduce. It was also key to the species' cognitive and emotional development. Development of the specifically human mind began in the playful exchange of gestures and facial expressions between human infants and their kin and non-kin caretakers. This exchange, Gallagher notes, "brings the infant into a direct relation with another person and starts them on a course of social interaction."[46] We do not develop singly but within "the larger system of body-environment-intersubjectivity."[47] This back-and-forth yields a "unified common intersubjective space"[48] with a wide variety of others that even infants know are different from themselves. It is not an undifferentiated we-space but an I-You space.[49] Each stage of human cognitive and emotional growth emerges from this interaction to arrive at what Sarah Hrdy calls "emotional modernity,"[50] the capacities to grasp and coordinate with (1) the attention of others, (2) the intention of others, and (3) the emotions of others in order to sustain relationships through which one feels safe and learns about the world. Importantly, learning and relating generalize non-kin strangers.

Michael Tomasello's work on cognitive development adds that joint attention and intention created the basis for role reversal and recursive thinking. Role reversal entails understanding, for instance, that if I touch your arm, you touch not your arm but *my* arm; it's touching the arm of the *other* that is the task. Role reversal allows tasks to be separated from the actor and to be distributed to various persons. Recursive thinking involves understanding that the other person wants me to know that she knows that I know, etc. Together, these allow for complex, collaborative endeavors where each knows what the other's role is and, importantly, trusts that the other will do it. Even before *H. sapiens*, Bellah notes, the *H. erectus* evolved "an entirely new level of social organization beyond anything seen in nonhuman primates."[51] In Tomasello's words, "The key novelties in human evolution were . . . adaptations for an especially cooperative, indeed hypercooperative, way of life."[52]

In sum, interactive exchange bridges otherness. It emerges from and reinforces our hypercooperativity. "It isn't just," Alison Gopnik concludes, "that without mothering, humans would lack nurturance, warmth, and emotional security. They would also lack culture, history, morality, science, and literature."

RELATIONALITY, COOPERATIVITY, AGGRESSION INTRA-GROUP

In addition to the psychological argument, biology too notes that *H. sapiens* evolved toward "hypercooperativity" and "reciprocal altruism."[53] "Overall," Richard Wrangham notes, "physical aggression in humans happens at less than 1 percent of the frequency among either of our closest ape relatives . . . we really are a dramatically peaceful species."[54] Benefits of cooperativity included improved food gathering, protection from animal predators, and other collaborative projects as well as more equitable resource distribution yielding greater longevity for more people and thus greater chances at reproduction. Kappeler et al. add that,

> individuals characterised by above-average frequencies of affinity, affiliation and mutual support, which are said to have strong social bonds, enjoy greater reproductive success, higher infant survival and greater longevity, and these effects are independent of dominance rank.[55]

"Natural selection," Robert Seyfarth and Dorothy Cheney similarly write, "therefore appears to have favored individuals who are motivated to form long-term bonds *per se* not just bonds with kin."[56] Frans de Waal in turn observes, "We owe our sense of fairness to a long history of mutualistic cooperation," again, not just with kin.[57] When Donald Pfaff writes that we are "wired for goodwill,"[58] he is not suggesting an absence of all competition and aggression among hunter-gatherers. Rather, he recognizes that episodic aggression occurred *amid* evolutionarily selected egalitarianism and cooperativity because the latter two were significantly advantageous within primary groups and often between groups, as well.

INTER-GROUP AGGRESSION

If intra-group cooperativity is high, it might be argued that aggression is more frequent inter-group owing to less need for cooperation and thus a lower bar to violence. Inter-group aggression ranges from one-on-one intimidation to raiding and war. Among hunter-gatherers, such aggression was episodic, and dependent on (1) rewards being sufficient to justify risks, (2) chances of success being high, and (3) risk of harm to oneself being low.[59] While low-risk raiding opportunities presented themselves, among hunter-gatherers, where stored goods were negligible, the risk-benefit analysis did not come out in favor of raiding consistently enough for raids to become systemic practice.

Indeed, among Pleistocene hunter-gatherers, food shortages may have led to cooperation. If, in a simple example, hunter-gatherer bands battle each other to be the sole group to hunt an animal, the winner may end with more food. But many will be downed in the inter-band fight, the capacity to overpower the animal will be diminished, and chances increase of becoming the animal's meal rather than making it one's own. Cooperation may be the better survival strategy as more people live (and may later reproduce) and chances of succeeding in the hunt rise. Moreover, the *value* of cooperation and food sharing becomes part of the modus vivendi in this long, 250,000 or so years of human development. Similarly, if one group raids the food cache of another, chances of retaliation are not trivial—not only with the motive of hunger but with added anger at the initial attack. Cooperation or at least non-engagement may be the more productive route. In both cases, "parochial altruism," concern for the in-group, led to *non*-aggressive strategies between groups.

In sum, David Barash finds that war is not genetically hard-wired but rather "historically recent," "erratic in worldwide distribution," and "a capacity." Capacities are "derivative traits that are unlikely to have been directly selected for but have developed through cultural processes . . . capacities are neither universal nor mandatory."[60] R. Brian Ferguson,

Douglas Fry, Gary Schober, Kai Bjorkqvist, and Patrik Soderberg, among others, make a similar case that *systemic* raiding and war required specific ecologies and conditions not found among hunter-gatherers. Indeed, Clare et al. find "no conclusive evidence for intergroup fighting in the early Pre-Pottery Neolithic" and warn of the "'bellicosification' of prehistory."[61] Importantly, while we find, in fossil material before 8,000 BCE, evidence of cut marks on bones, arrowheads embedded in the body, and other marks of trauma, little can be identified as systemic inter-group aggression. Kissel and Kim, in their important literature review, note, "Such signatures alone are insufficient to indicate violence, much less organized violence, between groups."[62] Kissel and Kim agree with Keeley[63] and Fry, Schober, and Bjorkqvist that periods of the Holocene show "virtually no signs of violent conflict" intergroup, much less intra-group.[64]

Finally, Kissel and Kim note that evidence of coalitional aggression (organized raiding and war) prior to 8,000 BCE, such as that cited by Steven Pinker,

> overlooks much of the evolutionary pressures that affected our ancestors. Evidence from Nataruk, Jebel Sahaba, and other cemetery burials demonstrate violence, and perhaps collective violence. However, anthropologists need to be clear that this represents only a tiny portion of the human evolutionary record.

THE EMERGENCE OF SEVERE, SYSTEMIC AGGRESSION

With hypercooperativity as the hunter-gatherer modus vivendi prior to 8,000 BCE, what was responsible for the shift to the systemic practice of severe aggression found after that date? Severe, systemic aggression includes endemic raiding and warfare, the enslavement of captive populations, and the subjection of domestic populations to maiming, torture, imprisonment, impoverishment, enslavement, and conspecific killing (killing within the species).

One understanding of the shift looks at the effects of sedentarism and agriculture, among the most significant changes in human development. They allowed for regular surpluses ever-present as lures to plunder, which in turn led to resource monopolizability and the development of inequality and sociopolitical hierarchies. With the new agrarian surpluses, the potential rewards of stealing by force, both intra- and inter-group, outweighed the risks far more often than they had under hunter-gatherer surplus-less mobility. "Hunters and gatherers," Kappeler explains, "forage cooperatively, share what they hunt/collect, and consume it on the spot. Agriculturalists don't rely on cooperation; they produce surplus stock for themselves which can be taken by force." Fry's large-scale study on present-day foragers, though limited in applicability to the Pleistocene, found that non-egalitarian societies engaged in warlike activity while the majority of (egalitarian) mobile foragers did not. Fry posits that the accumulation of stored goods and development of hierarchies in non-egalitarian societies meaningfully increase the likelihood of raiding or warfare.

The desire to grab what others have and the need to constrain those wanting to grab one's own cache was a first prod to endemic inter- and intra-group violence. "A tiny ruling group that used coercive powers to augment its authority," Bellah writes, "was sustained by agricultural surpluses and labor systematically appropriated from a much larger number of agricultural producers."[65] A second prod to aggression, van Schaik and Michel note, was the resentment that emerged as coercive, monopolizing hierarches violated evolution-bred cooperativity. Thus, protest and resistance added another layer of societal aggression to the monopolization of resources. Bellah describes a third prod in the lure not of goods but of politico-military power. While the first monopolizers grab resources, the next monopolizer has two things to grab: resources and the elite position in the hierarchy that the first monopolizer has. Bellah writes, "Large, prosperous societies are almost always in danger from the havenots at their fringe, or from other prosperous groups who would like to become even more prosperous. In a situation of endemic warfare, the successful warrior emanates a sense of mana or charisma, and can use it to establish a following" to take as much power and materièl as possible.[66]

In sum, the manifold, radical changes that brought inequality and hierarchy to agrarian living may have been sufficient to violate longstanding hyper-cooperativity—to turn episodic aggression amid prevailing cooperation into systemic, violent practice.

CONCLUSION

If hypercooperativity with robust, mutual fairness and sharing norms was evolutionarily selected for 95 percent of human evolution, if—changing discourses—relationality as reciprocal relations among distinct persons is the human ontology or baseline, we will flourish to the extent that we see to this cooperativity and relationality in the way we structure our modes of living. Should we not, we risk going against evolution and the structure of existence. And little good can come of that.

The emerging problems are two: on one hand, undue situatedness-in-relation, untempered by distinction, yields what Luigino Bruni calls the group as "gigantic I"[67]—both oppressive top-down control and stultifying conformity riddled with prejudice. Within groups, such situatedness is a pretext to stanch political or socioeconomic change and a club for those who want to keep others—women, minorities—out of the club. It has dire consequences for freedom, innovation, and wealth creation. Between groups, it yields zero-sum calculations, "us vs. them" thinking, and often violence.

On the other hand, undue distinction, separability from relation, brings greed, self-absorption, abandonment, and anomie. Persistent focus on the separate self—on the *exit*, on evasion of reciprocal responsibility—yields what Charles Taylor and Glen Stassen call the buffered self[68] and Luke Bretherton calls "isolated choosers"[69] self-absorbedly concerned with "me, my firm, my portfolio"—as dramatized in tragi-comic, cinematic reflection: *The Wolf of Wall Street* (1929), *Wall Street* (1987), *Wall Street Warriors* (2006), *Margin Call* (2011), *Wall Street: Money Never Sleeps* (2010), and *The Wolf of Wall Street* (2013), all explorations of the culture of self-interested avarice. In addition is the insidiousness of anomie. Undue emphasis on the separate person may leave one not freely flourishing but isolated and unmoored. Able to choose but with few choices that inspire or give meaning and energy to life, one becomes *not unsatisfied but unsatisfiable* and vulnerable to "deaths of despair."[70]

Given our relational nature, we cannot develop priorities and purposes on our own, and even if that were possible, we would lack the community networks and governmental policies and institutions to realize them. Yet in a society of undue separability, such support too is undermined, for with excessive separability comes also a fraught view of government. In a culture of exit, government, the largest agent of common effort, is a priori suspect, and so too its educational, health care, or economic programs that give citizens a leg up. As the enforcer of common responsibilities (taxes, labor and consumer protection, market and environmental regulation, etc.), it is seen as the foe of individual freedom. Contempt for government becomes the political standard on which governmental programs for the common good must justify themselves.

Yet perhaps our long experience of hypercooperativity remains with us as a resource for greater cooperativity today—at least more so than if humanity had never lived in cooperative conditions. While there is a substantial library of economic proposals making the argument for a more relational economics,[71] it is not much implemented in some measure owing to insufficient grasp of relationality and thus insufficient popular and political will. Thus, "while structural reforms may well be necessary," Mary Hirschfeld writes, "the analysis suggests that we need to work on shifting the culture."[72] We must adjust the lenses through which we see the world toward the relationality that grounds and governs it. Though he is now touted as the guru of greed, Adam Smith understood the role of relationality in economics: in markets as in all of society, he wrote, each should "endeavor, as much as he can, to put himself in the situation of the other, and to bring home to himself every little circumstance of distress which can possibly occur to the sufferer."[73]

NOTES

1. Darcia Narvaez, *Neurobiology and the Development of Human Morality: Evolution, Culture, and Wisdom* (New York: W. W. Norton, 2014), 438.
2. Yoram Hazony, *The Philosophy of Hebrew Scripture* (Cambridge: Cambridge University Press, 2012), 22, 249.
3. Ian Barbour, *When Science Meets Religion: Enemies, Strangers, or Partners?* (San Francisco: Harper San Francisco, 2000), 164.
4. Elliot R. Wolfson, *Giving Beyond the Gift: Apophasis and Overcoming Theomania* (New York: Fordham University Press, 2014), Kindle ed., Kindle loc. 2812.
5. Thomas Aquinas, *Summa Theologica 1–5*, trans. Fathers of the Dominican English Province (Louisville, Ky.: Westminster John Knox Press, Christian Classics, 1948), Ia, q. 105, art. 5.
6. Catherine Keller, *Cloud of the Impossible: Negative Theology and Planetary Entanglement* (New York: Columbia University Press, 2014), 56.
7. Thomas Aquinas, *Summa Theologica* I, q. 47, a. 3.
8. Mary Hirschfeld, *Aquinas and the Market* (Cambridge, Mass.: Harvard University Press, 2018), 92 (emphasis mine).
9. Emmanuel Levinas, *Beyond the Verse: Talmudic Readings and Lectures*, trans. Gary D. Mole (Bloomington: Indiana University Press, 1994), 142.

10. Martin Buber, *Between Man and Man* (New York: Routledge, 1993), 203.
11. John Zizioulas, "Human Capacity and Incapacity: A Theological Exploration of Personhood," *Scottish Journal of Theology* 28 (1975): 409.
12. Elisabeth Moltmann-Wendel, *I Am My Body: A Theology of Embodiment* (New York: Continuum, 1995), 43.
13. Kirk Wegter-McNelly, *The Entangled God: Divine Relationality and Quantum Physics* (New York: Routledge), 136.
14. Karl Rahner, *Foundations of Christian Faith: An Introduction to the Idea of Christianity*, trans. William V. Dych (New York: Seabury, 1978), 15, 17, 62, 447, 456.
15. Catherine Keller and Laurel C. Schneider, eds., *Polydoxy: Theology of Multiplicity and Relation* (New York: Routledge, 2012), 7.
16. Marinus H. van Ijzendoorn, Jesús Palacios, Edmund J. S. Sonuga-Barke, Megan R. Gunnar, Panayiota Vorria, Robert B. McCall, Lucy Le Mare, Marian J. Bakermans-Kranenburg, Natasha A. Dobrova-Krol, and Femmie Juffer, "Children in Institutional Care: Delayed Development and Resilience," *Monographs of the Society for Research in Child Development* 76, no. 4 (2011): 8–30.
17. Matthew Pantell, David H. Rehkopf, Douglas Jutte, S. Leonard Syme, John Balmes, and Nancy Adler, "Social Isolation: A Predictor of Mortality Comparable to Traditional Clinical Risk Factors," *American Journal of Public Health* 103, no. 11 (2013): 2056–62.
18. John Cacioppo and Stephanie Cacioppo, "Social Relationships and Health: The Toxic Effects of Perceived Social Isolation," *Social and Personality Psychology Compass* 8, no. 2 (2014): 58–72; Nicholas Leigh-Hunt, David Baguley, Kristin Bash, Victoria Turner, Stephen Turnbull, Nicole K. Valtorta, and Woody Caan, "An Overview of Systematic Reviews on the Public Health Consequences of Social Isolation and Loneliness," *Public Health* 152 (2017): 157–71.
19. Gregory of Nyssa, "Ad petrium," in *Saint Basil: The Letters*, vol. II, trans. Roy J. Deferrari (London: Loeb Classical Library Heinemann, 1961), Ep. 207–9.
20. Edith Stein, *Endliches und Ewiges Sein* (Leuven: Nauwelaerts; Freiburg: Herder, 1950), 324.
21. John Milbank, "Christianity and Platonism in East and West," in *Divine Essence and Divine Energies: Ecumenical Reflections on the Presence of God in Eastern Orthodoxy*, ed. Constantinos Athanasopoulos and Christoph Schneider (Cambridge, Mass.: James Clarke, 2013), 177.
22. Wegter-McNelly, *Entangled God*, 128–29; Wolfhart Pannenberg, *Systematic Theology*, trans. Geoffrey W. Bromiley (Grand Rapids, Mich.: Eerdmans, 1991), 1:298, 323, 335 (emphasis mine).
23. Stanley Grenz has described the recent inquiry into the triune God as "the final touchstone for speaking about human personhood" and human relations;

Grenz, *The Social God and the Relational Self* (Louisville: Westminster John Knox Press, 2001), 57.
24. See, for instance, Aquinas, *Summa Theologica*, Ia, q. 40, art. 2.
25. Wolfhart Pannenberg, *Jesus-God and Man*, trans. Lewis Wilkins and Duane Priebe (Philadelphia: Westminster, 1977).
26. Jürgen Moltmann, *Ethics of Hope*, trans. Margaret Kohl (Minneapolis: Fortress Press, 2012), 220.
27. Daniel Bell Jr., "A Theopolitical Ontology of Judgment," in *Theology and the Political: The New Debate*, ed. Creston Davis, John Milbank, and Slavoj Žižek (Durham, N.C. and London: Duke University Press, 2005), 221.
28. Jean Lee, "The Two Pillars Paradigm: Covenant as a Relational Concept in Response to the Contract-Based Economic Market" (PhD diss., University of Edinburgh, 2010), 59.
29. Stephen Geller, "The God of the Covenant," in *One God or Many? Concepts of Divinity in the Ancient World*, ed. Barbara N. Porter, Transactions of the Casco Bay Assyriological Institute 1 (Casco Bay, Maine: Casco Bay Assyriological Institute, 2000), 295–96.
30. Lenn E. Goodman, *On Justice: An Essay in Jewish Philosophy* (New Haven: Yale University Press, 1991), 41–42.
31. Robert Bellah, *Religion in Human Evolution* (Cambridge, Mass.: Belknap Press of Harvard University Press, 2011), Kindle ed., Kindle loc. 4864.
32. Marcel Mauss, *The Gift: The Form and Reason for Exchange in Archaic Society*, trans. W. D. Halls (London: Routledge, 1990); Lewis Hyde, *The Gift: Imagination and the Erotic Life of Property* (New York: Vintage Books, 1983).
33. Eric Mount Jr., *Covenant, Community and the Common Good: An Interpretation of Christian Ethics* (Cleveland, Ohio: Pilgrim Press, 1999), 1.
34. Amos 5:21–24: "I hate, I despise your religious festivals; your assemblies are a stench to me. . . . But let justice roll on like a river, righteousness like a never-failing stream"; Proverbs 21:3: "To do what is right and just is more acceptable to the Lord than [ritual] sacrifice."
35. See also Jacques Godbout and Alain Caille, *The World of the Gift*, trans. Donald Winkler (Montreal: McGill-Queen's University Press, 1998).
36. See John Milbank, "Can the Gift Be Given? Prolegomena to Any Future Trinitarian Metaphysic," *Modern Theology* 11, no. 1 (1995): 119–61.
37. Marcia Pally, "Sacrifice amid Covenant," in *Mimesis and Sacrifice*, ed. Marcia Pally (London: Bloomsbury Academic), 108–9.
38. Irenaeus, *Against Heresies [Adversus haereses]*, ed. Alexander Roberts and James Donaldson (Grand Rapids, Mich.: Eerdmans, 1989; Latin original 180 CE), 478.
39. Levinas, *Beyond the Verse*, 142.

40. Emanuel Levinas, *In the Time of Nations*, trans. Michael B. Smith (London: Athlone Press, 1994), 171.
41. Emmanuel Levinas, "Revelation in the Jewish Tradition," in *Beyond the Verse*, 146–47.
42. Richard Kearney, "Paul's Notion of Dunamis: Between the Possible and the Impossible," in *St. Paul Among the Philosophers*, ed. John Caputo and Linda Alcoff (Bloomington: Indiana University Press, 2009), 142–59, especially 143, 155.
43. Frans B. M. de Waal, "Primates–A Natural Heritage of Conflict Resolution," *Nature* 289 (2000): 589; Michael Tomasello, *Becoming Human: A Theory of Ontogeny* (Cambridge, Mass.: Harvard University Press, 2019), Kindle ed., Kindle locs. 5521–22.
44. Peter Kappeler, "A Comparative Evolutionary Perspective on Sacrifice and Cooperation," in *Mimesis and Sacrifice: Applying Girard's Mimetic Theory Across the Disciplines*, ed. Marcia Pally (London: Bloomsbury Academic, 2019), 39.
45. Christopher Boehm, *Moral Origins: The Evolution of Virtue, Altruism, and Shame* (New York: Basic Books), 161.
46. Shaun Gallagher, *How the Body Shapes the Mind* (Oxford: Oxford University Press, 2005), 128; see also 224–25, 244–45.
47. Gallagher, *How the Body Shapes the Mind*, 242–43.
48. Vittorio Gallese, "'Being Like Me': Self-Other Identity, Mirror Neurons, and Empathy," in *Perspectives on Imitation*, ed. Susan Hurley and Nick Chater (Cambridge, Mass.: MIT Press, 2005), 105, 111.
49. Vasudevi Reddy, *How Infants Know Minds* (Cambridge, Mass.: Harvard University Press, 2008), 19–21; R. Peter Hobson and Jessica A. Hobson, "Joint Attention or Joint Engagement?," in *Joint Attention*, ed. Axel Seemann (Cambridge, Mass.: MIT Press, 2012), 120–21.
50. Sarah B. Hrdy, *Mothers and Others: Evolutionary Origins of Mutual Understanding* (Cambridge, Mass.: Harvard University Press), 204–6, 282.
51. Bellah, *Religion in Human Evolution*, Kindle loc. 2019.
52. Tomasello, *Becoming Human*, Kindle locs. 5521–22.
53. Robert L. Trivers, "The Evolution of Reciprocal Altruism," *Quarterly Review of Biology* 46, no. 1 (1971): 35–37.
54. Richard Wrangham, *The Goodness Paradox* (New York: Knopf Doubleday, 2019), 19.
55. Peter Kappeler, Claudia Fichtel, and Peter van Schaik, "There Ought to Be Roots: Evolutionary Precursors of Social Norms and Conventions in Non-Human Primates," in *The Normative Animal?: On the Anthropological Significance of Social, Moral, and Linguistic Norms*, ed. Neil Roughley and Kurt Bayertz (Oxford: Oxford University Press, 2019), 65–82; see also Oliver Schülke and Julia

Ostner, "Ecological and Social Influences on Sociality," in *The Evolution of Primate Societies*, ed. John C. Mitani, Ryne A. Palombit, Peter M. Kappeler, Joseph Call, and Joan B. Silk (Chicago: University of Chicago Press, 2012), 195–219; Joan B. Silk, "The Adaptive Value of Sociality in Mammalian Groups," *Philosophical Transactions of the Royal Society B* 362 (2007): 539–59.

56. Robert Seyfarth and Dorothy Cheney, "The Evolutionary Origins of Friendship," *Annual Review of Psychology* 63 (2012): 170.
57. Frans B. M. de Waal, "One for All," *Scientific American* 311 (2014): 71; see also Samuel Bowles and Herbert Gintis, *A Cooperative Species: Human Reciprocity and Its Evolution* (Princeton: Princeton University Press, 2013); Sarah Brosnan and Frans B. M. de Waal, "Evolution of Responses to (Un)Fairness," *Science* 346 (2014): 1251776-1—7; Joan B. Silk and Bailey R. House, "Evolutionary Foundations of Human Prosocial Sentiments," *Proceedings of the National Academy of Sciences* 108, no. S2 (2011): 10910–17.
58. Donald W. Pfaff, *The Altruistic Brain: How We Are Naturally Good* (New York: Oxford University Press, 2014).
59. Wrangham, *Goodness Paradox*, 262.
60. David Barash, "Is There a War Instinct?" *Aeon*, September 19, 2013, http://aeon.co/magazine/society/human-beings-do-not-have-an-instinct-for-war (accessed May 18, 2019).
61. Lee Clare, Oliver Dietrich, Julia Gresky, Jens Notroff, Joris Peters, and Nadja Pöllath, "Ritual Practices and Conflict Mitigation at Early Neolithic Körtik Tepe and Göbekli Tepe, Upper Mesopotamia: A Mimetic Theoretical Approach," in *Violence and the Sacred in the Ancient Near East: Girardian Conversations at Çatalhöyük*, ed. Ian Hodder (London: Cambridge University Press, 2019), 101.
62. Marc Kissel and Nam C. Kim, "The Emergence of Human Warfare: Current Perspectives," *American Journal of Physical Anthropology* 168, no. S67 (2019): 151.
63. Lawrence Keeley, "War before Civilization—15 Years On," in *The Evolution of Violence*, ed. Todd Shackelford and Ranald Hansen (New York: Springer, 2014), 30.
64. Kissel and Kim, "Emergence of Human Warfare," 155.
65. Bellah, *Religion in Human Evolution*, Kindle locs. 3279–81.
66. Bellah, *Religion in Human Evolution*, Kindle locs. 3974–76.
67. Luigino Bruni, *The Wound and the Blessing: Economics, Relationships and Happiness*, trans. N. Michael Brennen (Hyde Park, N.Y.: New City Press, 2012), 59.
68. Glen Stassen, *A Thicker Jesus: Incarnational Discipleship in a Secular Age* (Louisville: John Knox Press, 2012), 101; Charles Taylor, "Buffered and Porous Selves," *The Immanent Frame*, September 2, 2008, http://blogs.ssrc.org/tif/2008/09/02/buffered-and-porous-selves/.

69. Luke Bretherton, *Hospitality as Holiness: Christian Witness amid Moral Diversity* (Farnham: Ashgate, 2006), 16.
70. Anne Case and Angus Deaton, "Mortality and Morbidity in the 21st Century," *Brookings Papers on Economic Activity* (2017): 397–476; United States Congress, Joint Economic Committee, *Long-Term Trends in Deaths of Despair*, by Scott Winship, September 5, 2019, https://www.jec.senate.gov/public/index.cfm/republicans/2019/9/long-term-trends-in-deaths-of-despair (accessed February 9, 2020).
71. Thomas Piketty, *Capital in the Twenty-First Century*, trans. Arthur Goldhammer (Cambridge, Mass.: Harvard University Press, 2014); Luigino Bruni and Stefano Zamagni, *Civil Economy: Efficiency, Equity, Public Happiness* (Bern: Peter Lang, 2007); Duncan K. Foley, *Adam's Fallacy: A Guide to Economic Theology* (Cambridge, Mass.: Harvard University Press, 2006); Joseph Stiglitz, *The Price of Inequality: How Today's Divided Society Endangers Our Future* (New York: Norton, 2013), among others.
72. Mary Hirschfeld, "Rethinking Economic Inequality: A Theological Perspective," *Journal of Religious Ethics* 47, no. 2 (2019): 278.
73. Adam Smith, *The Theory of Moral Sentiments*, ed. David Raphael and Alec Macfie (Oxford: Clarendon Press, 1759), 21.

In Whose Interest? Matthew 25:14–30 as a Theo-Economic Parable Hard at Work

HILARY MCKANE

Financial and economic language pervade the teachings, images, and metaphors found in the New Testament and deployed in Christian theology. While much scholarly attention has been paid to language and teachings about wealth and poverty, there has been much less work done to examine the ways that financial and economic language have provided structure for the teachings, images, and theological concepts that we find in biblical texts.[1] This chapter looks at the "parable of the talents" found in Matthew 25:14–30 to examine both how financial systems informed the creation of the parable and its attendant theological implications, as well as how the parable then functioned historically to support those same systems. Although this parable has traditionally been interpreted as an allegorical exhortation to Christians encouraging bold action and risk taking as necessary in pursuit of the kingdom of God, its reliance on the language and structures of money, lending, banking, and earning interest points to more complex entanglements among the economic, political, and theological realms. The ethics of using money to make money and accruing huge sums of wealth at the expense of others are at the heart of the parable found in Matthew 25:14–30.[2] This chapter seeks to press on those tensions while asking more critical questions about how the parable functions. What theological work is this parable doing?[3] And whose interests does it serve?

Historically, the parable of the entrusted money was used as evidence in debates about lending money at interest and the ethics of usury for Christians. This chapter will examine the reception history of the parable at critical moments of economic transition to articulate connections between the economic context of interpreters and their readings of the parable. Beginning with Origen's early and influential reading, I will then trace the ways that the parable is interpreted in the twelfth and thirteenth centuries by Thomas Aquinas and others in the context of a rapidly expanding money economy. Martin Luther and John Calvin's discussions of moneylending within the context of not only the Reformation, but also social and economic upheavals, provide the next touchpoint in the reception history of the parable. Finally, I will analyze the ways that several significant modern commentaries address the parable from within the context of North American global capitalism. This analytical framework highlights the impact of an interpreter's economic context on their interpretation while underscoring the interconnectedness of economics, parables, and theological reflection.

When the master in the parable returns "after a long time" to "settle accounts" with his slaves, he is pleased that the first two slaves have each returned what was entrusted to them with 100 percent interest; the third slave is castigated as wicked and lazy. If the master of the parable functions as a God-figure (and this should not be taken for granted as a given), then the parable seems to hold moneylending and the accrual of exorbitant interest in very high esteem. God thus functions metaphorically as a proto-venture capitalist,[4] entrusting, and indeed expecting, that money invested will always make more money. The consequences of not earning enough interest are very dire indeed. The use of moneylending and earning interest as conceptual frames for illustrating something about the relationship between God and human beings raises fascinating, and often troubling, questions. We should not be surprised that economic and political conceptual frames structure so many parables, given that parables draw on images and themes that structure everyday life and economic exchanges certainly undergird all of daily life, both in the first century and in our own time. However, we should be less interested in trying to determine whether the master does indeed function as a God-figure in this parable, and we should be more interested in parsing the theological implications of the employment of economic concepts, both in the parables themselves and in later interpretations. The genre of parable is also important to consider here. In his seminal work on Jesus's parables, Crossan argues that while myth creates world, parables subvert world,[5] and they are always "participatory—because provocative—pedagogy."⁶ Going even further, Herzog claims that the parables "were not earthly stories with heavenly meanings but earthy stories with heavy meanings," exposing how oppression and exploitation functioned in the world of their hearers.⁷ This chapter examines the economic exploitation being exposed by this parable, moving beyond the ways that the parable may have functioned as subversive speech within its original context and into an analysis of the implications of the parable for both theology and economics within the context of contemporary global capitalism.⁸ With the subversive power of parables in mind, an examination of their connections to economic imagery—and reality— invites further attention.

Given the centrality of money in the parable, the work of Devin Singh is particularly helpful for thinking about the historical interaction and mutual shaping of money and theology. He argues that "money lends its logic to the structuring of theology, [and] God-talk repays by offering its prestige and sacred power to the world of exchange."⁹ Thus, an examination of the parable of the talents must take into account not only the ways that ancient thinkers are drawing on imagery of money and money lending to make a theological point; it must also ask about the ways that the parable then functions to provide theological support for an economy centered on money, credit, and the expectation of earning interest.

EXAMINING THE PARABLE WITHIN ITS FIRST-CENTURY ECONOMIC CONTEXT

To begin, there are interesting layers of financial and economic language embedded in the parable itself, and there are also subtle, but note-worthy tensions between financial language and agricultural language at work within the parable. Each of the three slaves was given a certain amount of *talenta*, "to each according to his ability" (*kata tēn idian dunamin*; Matthew 25:15). Both the slave who received five talents and the slave who received two talents "gained" (*kerdēso*) more talents, five more and two more respectively. *Kerdēso* has well-attested financial connotations.¹⁰ When the slaves are called before the master to give an accounting, both slaves are described as having been "faithful" (*pistos*) over a few things, so they will be set over many things (25:20). In stark contrast, the third slave hides the talent that he was given in the ground (*en tē gē*), and his critical response to the master employs decidedly agricultural imagery. The third slave describes the master as a "harsh man, reaping where you did not sow, and gathering where you did not scatter seed" (25:25). After calling the third slave "wicked and lazy," the master repeats this agricultural language and then angrily contends that the slave should have entrusted his money with the bankers so that the master would have received what was his "with interest" (25:26–27). Finally, the master hurls the ultimate financial insult at the third slave: he is *achreion*, "worthless" (25:30).[11] Because the third slave does not increase the master's net worth, he himself is worthless, of no worth and having no value to the master. Thus, he is cast "into the outer darkness, where there will be weeping and gnashing of teeth" (25:30).

The parable centers on the issue of using money to make more money, and although this impetus seems familiar and perhaps even laudatory in a twenty-first-century American context, it would have resonated very differently in the context of a first-century Jewish audience under Roman occupation. The expectation of exorbitant profits by the master of the parable as well as his exhortation to the third slave that he should have at least invested the money with bankers raises questions about how such profits would have been achieved, and more specifically what role moneylending played in an ancient context. An examination of the rising levels of indebtedness not only sheds light on early responses to the parable; it also reminds us that the expanding money economy informs and infuses the language and imagery available for theological reflection. In his work on debt in the first century, Douglas Oakman argues that there is both direct and indirect evidence for escalating debt, including fiscal pressure, population pressure, and popular unrest.[12] An oft-cited example of a popular uprising with a specifically economic focus comes from Josephus, who describes a revolt in which the rebels "carried their combustibles to the public archives, eager to destroy the money-lenders bonds and to prevent the recovery of debts, in order to win over a host of grateful debtors and to cause a rising of the poor against the rich."[13] In an agrarian economy like that of Judea and Galilee in the first century, farmers sometimes needed loans to cover costs at the beginning of a growing season, but they were often forced to put up their land as collateral. If an extended drought or successive lean years meant that crops could not cover the repayment of the loan, then small landowners faced foreclosure.[14] Oakman notes that the result of escalating debt was "the growth of tenancy and the landless class" as "more and more land came under the control of fewer and fewer landowners."[15]

Another component of increasing indebtedness in the first century was the economic pressure that peasants experienced from Roman imperial exploitation. Paying taxes and tribute to Roman officials, particularly in an increasingly monetized economy, contributed to loss of land and the breakdown of traditional social and economic structures.[16] Horsley also asserts that after the Roman conquest, the "intense economic pressure on the peasant producers continued under the Roman governors and the Jewish high priests" such that "the people struggled under a double burden of taxation: tithes and other dues paid to the Temple and priests as well as tribute and other taxes to Rome."[17] Under this double burden, peasants hoped for a crop yield that would allow them to forestall foreclosure, even as they witnessed declining social and economic conditions and greater divisions between the very rich and the very poor. Amidst these conditions, the idea of using money to make money would have thrown into sharp relief the lived reality of peasants in contrast to the master's expectation of generating significant profits. For the vast majority of the population, who hovered near subsistence and struggled to survive,[18] an already rich man becoming inordinately richer through exploitative moneylending at the expense of the poor would have likely evoked admonition and anger, rather than admiration.

In addition to examining and contextualizing the role of moneylending in a first-century context, it is also instructive to examine the function of and attitudes about money itself in this context. These attitudes are also deeply connected to attitudes about labor. In an article that examines the parable of the talents, Richard Rohrbaugh describes the agrarian economy of the first century as one centered on the idea of "limited good" and draws helpful distinctions between use value and exchange value. In almost all peasant societies, the central elements of life—including land, wealth, honor, and status—are thought to exist in finite amounts, so that they can always be redistributed but they cannot be expanded. Essentially, this is an understanding of the world in which "the pie is limited" and where one person's increase necessitates another person's loss because "there is simply not enough of anything to go around or any way to increase the size of the pie."[19] Thus, from a peasant perspective, the master's acquisition of several additional talents would have necessarily come as a result of several others' significant losses. With regard to the functions of money, Rohrbaugh notes that peasants tend to think about money in terms of its use value; their labor yields products that can be consumed or, if needed, sold in order to get money that can then be used to buy other needed products.[20] However, merchants and moneylenders (and later capitalists) use money to buy things that they can sell again at a profit to acquire more money; money functions here primarily for its exchange value rather than for its use value. This was largely considered an unnatural use of money. Those who were using money to try to make more money were expecting money to become fertile or breed, assuming unnatural powers that it should not possess.[21] Thus, the parable seems to encapsulate everything that would have been abhorrent to peasants about the expanding monetary economy under Roman occupation: growing divisions between the rich and poor, exploitative lending practices, and the unnatural use of money to make more money.

Although a discussion about how to interpret the parable in light of these economic realities will take place later in the chapter, it is important to note at this juncture that the use of economic or financial imagery and concepts in the parables goes beyond simply drawing on images from everyday life to make theological points. On the one hand, some interpreters have argued that Jesus merely chose images that would have been familiar to his hearers to convey teachings about the kingdom of God, and so the images used and situations depicted in the parables do not actually speak to those realities in any meaningful way. In these kinds of readings, the parable of the talents does not intend to provide any reflection on wealth, moneylending, or economics. On the other hand, some interpreters have used the parable to make arguments about God's perspective on these very concerns. Historically, as we will see in the next section, discussions about usury often employed this parable as evidence both for and against its practice. Yet, beyond the question of whether the parables actively speak to economic realities, the inclusion of economic and financial images in the parables signals an entanglement between economics and theological reflection that runs much deeper than has been previously explored. On a fundamental level, this inclusion of economic concepts and images in the parables indicates the appropriateness of using these kinds of images to think with, and particularly to think about, God. Thus, even if they offer critiques of economic practices, these parables point to the difficulties of teasing apart economics and theology, and they implicitly endorse theo-economic reflection.

What is meant by theo-economic reflection? The term "theo-economic" first appeared in David Wilhite's work, where he urged resistance to the modern dichotomies between religion and economics.[22] In her work, Jennifer Quigley extends the framework for thinking about theo-economics, using it as "a methodological intervention that keeps to the fore the transactional entanglements of humans and divines and the possibility of divines as economically actant when we read early Christian texts."[23] For Quigley, one aspect of this intervention involves analyzing the ways that financial and economic language is used to make theological arguments.[24] It is this aspect of her methodology that I find particularly helpful. Although it might seem obvious that a parable focused on money would qualify as an example of using financial language to make a theological argument, my analysis uncovers additional layers of entanglements between economics and theology. The parable does not simply employ financial imagery to make a theological point, such that we can jettison the images of money and moneylending once we have uncovered the parable's point about, for example, the appropriate use of spiritual gifts. Instead, theo-economic reflection calls us to examine the ways that financial and economic realities structure the parable and inform its theological arguments while being attentive to the ways that the parable then informs theological reflection about economic practices. It is an attention to the "historical interaction and mutual shaping of fields" explored by Singh.[25] Theo-economic reflection on this parable, and on other parables with economic overtones, also involves attentiveness to the ways that financial and economic language and concepts explicitly and implicitly make their way into the work of interpreters.

CONTEXTUALIZING THE RECEPTION HISTORY OF THE PARABLE

The earliest extant commentary on the Gospel of Matthew comes from Origen, a scholar-teacher in Alexandria and then in Caesarea in the third century. Origen's father was martyred in 203 CE, when Origen was a teenager, and the family's property was also confiscated, leaving them in poverty.[26] A wealthy benefactor later supported Origen's education. Whether he was writing, teaching, or preaching, he focused on the interpretation of scripture, and in his work *De Principiis*, Origen argues that the Holy Spirit sometimes "hides the mysteries of God and salvation behind the stories in Scripture for those capable of searching out hidden meanings" and at other times creates the stories "such that those incapable of probing behind the texts for meaning might, nevertheless, be benefitted by reading the literal stories."[27] In Origen's commentary on the parable of the talents in Matthew 25, he moves quickly to a spiritualized allegorical reading of the elements of the parable. The master who goes out on a journey calls his servants[28] so that he can deliver to them "his approved money in pure sayings, since his 'sayings' are 'pure sayings,' silver that has been refined by fire, proven, cleansed seven times."[29] Thus, the money entrusted to each slave symbolizes the sayings or teachings of Jesus. The slaves are given different amounts of money based on their different levels of ability, and Origen asserts that this ability does not refer to any specifically financial ability, but rather refers to the ability to speak and benefit souls.[30] Even though Origen understands ability in this way, he goes on to describe the third slave as one who believes but chooses not to openly express one's faith—that is, one who guards what one has received "but they do not also add to it. They neither do business with it nor act openly in it."[31] Despite his attempt to spiritualize the elements of the parable, Origen slips back into the language of business and financial practices.

This slippage becomes even more evident as Origen justifies the treatment of the "wicked and lazy" third slave. He writes:

> We shall be condemned as wicked because we have not used that approved "money" of the words of the Lord, with which we could have done business and discussed the teachings of Christianity and acquired the higher mysteries of the goodness of God; and lazy because we have not worked with the teachings of God for the salvation either of ourselves or of others, when we ought "to have placed" our Lord's money, that is, his teachings, with pupils, the "money-changers" who test and prove all things so that they may hold the doctrine that is good and true, but reject what lacks value and is false, so that when the Lord comes, he may receive those teachings which we have brought together "with interest" and with what has been added by those who received them from us. For each one is "money," that is, every word that has the royal form of God and the image of his Word in itself, is legitimate. He will command the men, the "money-changers" who are able to pay back his interest, to believe the Word. And he who hastens to save another with himself just as he has been saved

pays back interest on the word, as Andrew did with Peter, Philip with Nathanael, and Matthew with the tax-collectors.[32]

If the teachings of God are the money, then Christians are exhorted to invest those teachings with the "moneychangers," who symbolize pupils in Origen's analysis. These faithful moneychangers are then able to receive this money ("teachings") and add to it "with interest" (other teachings). In the allegorical world of the parable, money refers to every word of God, and so every word of God is money. Thus, moneychangers are good and faithful students. These moneychangers "hold the doctrine that is good and true, but reject what lacks value and is false."[33] The language of value here is telling. Origen's interpretations of scripture, and especially his allegorical approach, wielded tremendous influence for centuries. Yet, the unacknowledged influence regarding this particular parable lies in his struggle to tease apart the spiritual and the economic. This struggle would become even more acute as later economic transitions and upheavals impacted interpretation.

In the thirteenth century, usury became one of the major issues discussed by Christian theologians, and the parable of the talents frequently found its way into their arguments. Although the Christian tradition had consistently condemned usury, economic shifts forced reevaluations of church doctrine on the issue. In the twelfth and thirteenth centuries, economic growth throughout Europe led to the spread of the monetary economy and the development of credit at an unprecedented scale, leading to a significant increase in the circulation and availability of currency.[34] Distinctions needed to be made between usury and interest, and questions about the legitimacy of making a profit also resurfaced. The church feared that society was beginning to be unsettled by the practice of usury, and at the Third Lateran Council in 1179 Pope Innocent IV expressed this fear with striking detail. He feared "that the countryside would be deserted, because peasants had become usurers or had been deprived of their cattle and tools by landowners, themselves attracted by the profits of usury"; in fact, the "appeal of usury threatened to depopulate the countryside and hamper agriculture and thus raised the spectre of famine."[35] The stakes could not be higher. Other condemnations of usury emerged during this period, including St. Anselm's assertions that the usurer is universally slothful and that usury constitutes theft.[36]

How could theologians reconcile their convictions that usury was forbidden by church teaching and the biblical texts with the parable's apparent endorsement of moneylending and the pursuit of exorbitant interest? Thomas Aquinas argued that usurious theft is a sin against justice and that money should not "reproduce itself" because its proper use is only in exchange and disbursement through ordinary transactions.[37] In his reading of the parable of the talents, Aquinas leans heavily on Origen and largely spiritualizes the talents in the parable, though there are moments where he cannot seem to avoid more literal financial readings. Aquinas claims that "the talents are different gifts of graces: for just as a weight of metal is called a talent, so grace is a weight because it inclines the soul."[38] These gifts are also "God's word and words of wisdom" because scripture often compares wisdom to riches.[39] When he analyzes the gains achieved by the first two slaves, Aquinas notes that "a man profits in two ways": he "profits for himself if he has understanding of Scripture" and "if he has charity he consequently profits for others."[40] The slippage between theology and economics becomes readily apparent as Aquinas deploys the language of profit to describe spiritual understanding as well as almsgiving. A similar slippage occurs when he attempts to explain why the third slave is characterized as "unprofitable"; the third slave "does not use [the good that he has] for others' benefit: for example, if he had understanding, he does not put it to good use by teaching others; and if he had money, he did not perform works of mercy."[41] Although he tries to adhere to a spiritualized interpretation of the parable that understands the talents entrusted to each slave as gifts of grace that should be used to benefit others, financial language and reflections about the proper use of money seep into Aquinas's analysis. In the centuries that follow, especially in the context of the Reformation, interpretations of the parable continue to spiritualize some elements of the parable while offering more explicit theological reflection on economic concerns.

In the midst of the Reformation and the economic upheavals that accompanied it, discussions of usury took on a decidedly different tone. As peasants pressed for social reforms, Martin Luther asserted that not only were Christians not obligated to observe "dead Mosaic ordinances," but that attempts to restructure society based on the injunctions of Moses and Jesus were utopian and even anti-Christian.[42] Germany in Luther's context "was in the throes of political and social upheaval . . . peasants and knights were already on the march . . . town and country alike groaned under the burden of monopoly prices and fixed interest charges on loans."[43] Luther actively worked to prevent a social revolution, believing instead that any reform should come from princes, not from peasants. John Calvin argued that usury was only wrong if it was excessive, constant, or taken from the poor.[44] Merchants and lawyers themselves were becoming more vocal during this period, writing pamphlets and advocating for their own interests in ways that undoubtedly influenced theological reflection. Thomas Wilson refers specifically to Luke's version of the parable of the talents in his 1572 *Discourse Upon Usury*, arguing that usury cannot be unequivocally condemned because God encouraged taking usury from strangers: "For is it not in S. Lukes ghospell that god said he would come and aske the money lent with usury, blaming him that did not put it for the gaine?"[45] As the economic necessity of lending money at interest became more intense, the parable of the talents seemed to indicate that Jesus was in no way prejudiced against bankers or interest-bearing loans.

Sir William Blackstone, a Protestant jurist (d. 1780), went so far as to claim that comprehensive bans of usury were remnants of the Dark Ages when commerce was stifled by tyranny and superstitions. For Blackstone, the "credit economy and its inseparable companion, the doctrine of loans upon interest, both owed their rebirth to the revival of true religion and real liberty in the Reformation."[46] As this credit economy continued to grow and expand its reach, the Deuteronomy passage that had long been used as the basis for the prohibition of usury seemed to also be a remnant of a distant past,[47] not something applicable to the contemporary economic context. The pseudonymous *Letters on Usury and Interest*, published in England in 1774, argues that one only needs to reread the parable of the talents to see that lending money at interest is useful, beneficial, and encouraged. The slippage between financial and theological language appears most tellingly in the last line of the following passage:

> The practice of lending money to interest is in this nation, and under this constitution, beneficial to all degrees; therefore it is beneficial to society. I say in this nation; which, as long as it continues to be a commercial one, must be chiefly supported by interest; for interest is the soul of credit and credit is the soul of commerce.[48]

Regulated usury becomes justified by necessity and utility to the extent that it becomes the soul of credit, which in turn signifies the soul of commerce. Amidst changing economic and financial landscapes, ancient texts are reinterpreted and pressed into service to support new ideological positions.

A THEO-ECONOMIC PARABLE HARD AT WORK

One of the ways that this parable has influenced theological reflection and its entanglements with economics can be seen especially clearly with regard to the word "talent" itself. In the first-century context of the parable,[49] the Greek word *talenton* referred to a specific denomination of currency: one talent was equivalent to about 6,000 denarii, or about twenty years' wages.[50] Yet the word "talent" emerged in English in the fifteenth century and referred to skill, aptitude, or any mental or physical abilities a person might have.[51] The modern English understanding of the word is based largely on historic interpretation of *this* parable. Thus, *talenton*—money—becomes talent—skill or ability—and subsequently talent as ability becomes commodified. Modern commentators often reinscribe this commodification as they navigate unconsciously between the entangled realms of theology and economy. Exhortations to risk-taking abound, exemplified by Arland Hultgren's assertion that "to be afraid or to refuse to use one's gift signifies failure."[52] He argues that the gifts of God are more like funds entrusted to people for a while, implying that God functions as the administrator of a divine trust fund. Those who hear the parable should internalize its encouragement to "get on with lives of self-abandon and witness, knowing that the grace of God in Christ will more than compensate for any mistakes they might make."[53] The slippage into the language of compensation here is quite telling. The money economy thoroughly informs the central structure of the parable, which in turn shapes theological claims about how God interacts with human beings, and these theological claims become worked out through the language of money.

When venture capitalism serves as a symbol for the work of faith in action, it becomes easier to judge those who hover near subsistence—they, too, are unwilling to take risks with what they have been given, and so they deserve to suffer the consequences of their (in)action. The parable seems to provide theological support for the adage that the rich get richer and the poor get poorer: "For to all those who have, more will be given, and they will have an abundance; but from those who have nothing, even what they have will be taken away" (Mt 25:29). It is at this point that we need to ask the pointed question: is greed good? Can this parable be good news to anyone other than the billionaires of the world?

Several scholars have drawn attention to the third slave's actions, reading against the grain of the parable to interpret his actions in more positive ways. In these readings, the third slave chooses to bury his one talent in the ground, not because he is inherently lazy, but because he refuses to participate in a system that exploits those who are already struggling.[54] He knows that the master expects much more than the legally sanctioned 12 percent interest rate.[55] Rather than using his money to accrue more money by extracting usurious interest from other desperate people, he steps outside the system altogether, keeping his one talent safe to return to the master. When confronted, he holds up a mirror to the money economy itself and to all those who unquestioningly perpetuate it. The pursuit of exorbitant profits becomes exposed for its callous disregard for the well-being of others or for any sense of the common good. Yet, even these readings can become problematic when they fail to push back against another exploitative aspect of the system that forms a central part of the parable—slavery as the commodification of human lives. Just as we need to take seriously the entanglements between economy and theology as we read the parables, we also need to address the entanglements between theology and the master/slave paradigm. To read the parable against itself and uphold the third slave as a model for refusing to participate in exploitative systems, we must also acknowledge that slaves were much more than metaphors and exemplars,[56] and their use (and often abuse) in parables to illustrate theological points were often employed to support slavery and had dangerous consequences throughout the history of their interpretation.

So, how can a parable composed in the first century help us construct a sustainable future in the twenty-first century, as economy, ecology, and democracy seem to be disassembling all around us? In conversation with Kathryn Tanner's attempt to provide a "Protestant anti-work ethic,"[57] the parable can be read as a cautionary tale against the unbounded pursuit of monetary interest precisely because it challenges us to confront the uncomfortable connections between faith in divine sovereignty and faith in the money market. More specifically, the parable forces us to grapple with the commodification of one's abilities and the fusion of productive capacity with personal worth.[58] Thus, we can begin to excavate the theo-economic work that the parable does in order to serve different interests—namely, our interest in creating a more just economic system that refuses to equate wealth with worth. The third slave serves as a model in that he refuses to participate in and openly critiques the exploitation of others, even at the risk of his own life. Perhaps the parable's exhortation to risk-taking remains poignantly present, but not in the traditional readings that allegorized and separated theology and economy. To construct a future that works for everyone, we must uncouple our faith in divine sovereignty and our faith in the sovereignty of the money market, critiquing and dismantling exploitation wherever we see it, regardless of what it costs us.

We need to intentionally interrupt the interpretive feedback loop constructed in and through this parable, in which the image of a harsh master who expects exorbitant interest on his investments becomes an image of God, the divine proto-venture capitalist, which is then marshaled to provide theological support for contemporary capitalism and all of its excesses. The commodification of abilities, which begins linguistically when the Greek *talenta* of this parable become the English talents (read as ability, skill, or aptitude), must be excised from our theology, precisely because it results in the commodification of human beings, as slaves or in other insidious forms of labor exploitation. Instead, the figure of the third slave points to the ways that we can read the parable against itself, holding up a mirror to the parable's apparent exhortation to greed and exploitation and exposing its (theo)logical manipulation and abuse.

Theo-economic reflection presses us to examine not only the connections between economic practices and theological concepts; it also necessitates that we pay attention to how the work of interpreters is influenced by their economic context. With attention and interest in the ways the financial language seeps into "spiritual" readings and theological reflection, we may see that the entanglements between economics and biblical studies run much more deeply than we could have ever imagined. Instead of seeking to offer a theological reflection on the parable of the talents or any of the parables, completely divorced from reflections about economic and financial practices, we need to press into these entanglements with more vigor in recognition of the impossibility of ever teasing apart economics and money from the other aspects of lived experience. Only then can we begin to see clearly who benefits from particular interpretations and modes of analysis and whose interests are served.

NOTES

1. Notable recent work in this area includes M. Douglas Meeks, *God the Economist: The Doctrine of God and Political Economy* (Minneapolis: Fortress Press, 1989); Devin Singh, *Divine Currency: The Theological Power of Money in the West* (Stanford: Stanford University Press, 2018); Jennifer A. Quigley, *Divine Accounting: Theo-Economics in the Letter to the Philippians* (New Haven: Yale University Press, 2021).
2. Another version of the parable is found in Luke 19:12–27. I have chosen to focus only on the version found in Matthew because Luke's version seems to include elements from another parable that complicate its structure; however, the central elements of the parable analyzed here—a wealthy man entrusting money to three slaves, the first two slaves using the money to earn more money, and the third slave returning the same amount of money he was given—are present in both versions, and so the analysis could be applied to both.
3. Here I am thinking with Vincent Wimbush and his work toward moving beyond exegesis toward excavating the scriptures, specifically excavating the work that we make scriptures do for us; see Wimbush, "Introduction: TEXTures, Gestures, Power," in *Theorizing Scriptures: New Critical Orientations to a Cultural Phenomenon*, ed. Vincent Wimbush (New Brunswick, N.J.: Rutgers University Press, 2008), 1–16.
4. See William R. Herzog II, *Parables as Subversive Speech: Jesus as Pedagogue of the Oppressed* (Louisville, Ky.: Westminster John Knox Press, 1994), 159.
5. John Dominic Crossan, *The Dark Interval: Towards a Theology of Story* (Sonoma, Calif.: Polebridge Press, 1988).
6. John Dominic Crossan, *The Power of Parable: How Fiction by Jesus Became Fiction about Jesus* (New York: HarperOne, 2012), 95.
7. Herzog, *Parables as Subversive Speech*, 3.
8. Thus, I am less interested in trying to discern or recover the "original" meaning of the parable than I am in exposing the ways that interpretation has been impacted by economic context and reflecting on how the parable can speak to theo-economic questions in the current context. Here I am thinking with questions that Elisabeth Schüssler Fiorenza raised in her 1987 Presidential Address to the Society of Biblical Literature: "The rhetorical understanding of discourse as creating a world of pluriform meanings and a pluralism of symbolic universes, raises the question of power. How is meaning constructed? Whose interests are served? What kind of worlds are envisioned? What roles, duties, and values are advocated? Which social-political practices are legitimated?" Schüssler Fiorenza goes on to discuss an ethics of historical reading that shifts from focusing on "what the text meant" to "what kind of readings can do justice to the text in its historical

contexts"; Fiorenza, "The Ethics of Biblical Interpretation: Decentering Biblical Scholarship," *Journal of Biblical Literature* 107, no. 1 (March 1988): 14.
9. Singh, *Divine Currency*, 2.
10. Quigley, *Divine Accounting*, 72.
11. The King James Version translates this word as "unprofitable," a perhaps even more pointed economic insult.
12. Douglas E. Oakman, *Jesus and the Peasants* (Eugene, Ore.: Cascade Books, 2008), 19.
13. Josephus, *Jewish War* 2.427, cited in Oakman, *Jesus and the Peasants*, 26.
14. Herzog argues that this was in fact the purpose of making such loans; they were not intended to yield significant profits in the short term but instead functioned as a means for elites to extract more and more land; Herzog, *Parables as Subversive Speech*, 161.
15. Oakman, *Jesus and the Peasants*, 25.
16. Richard A. Horsley, *Jesus and the Spiral of Violence: Popular Jewish Resistance in Roman Palestine* (San Francisco: Harper & Row, 1987), 11.
17. Horsley, *Jesus and the Spiral of Violence*, 13.
18. Steven J. Friesen, "Poverty in Pauline Studies: Beyond the So-Called New Consensus," *JSNT* 26, no. 3 (2004): 343.
19. Richard L. Rohrbaugh, "A Peasant Reading of the Parable of the Talents/Pounds: A Text of Terror?," *Biblical Theology Bulletin* 23 no. 1 (Spring 1993): 33.
20. Rohrbaugh, "Peasant Reading," 34.
21. Rohrbaugh, "Peasant Reading," 34.
22. David Wilhite, "Tertullian on Widows: A North African Appropriation of Pauline Household Economics," in *Engaging Economics: New Testament Scenarios and Early Christian Reception*, ed. Bruce W. Longenecker and Kelly D. Liebengood (Grand Rapids, Mich.: Eerdmans, 2009), 222–42.
23. Quigley, *Divine Accounting*, 4.
24. Quigley, *Divine Accounting*, 4.
25. Singh, *Divine Currency*, 17.
26. Ronald E. Heine, "Introduction," *The Commentary of Origen on the Gospel of St. Matthew* (Oxford: Oxford University Press, 2018), 2.
27. Heine, "Introduction," 12.
28. This translation of Origen renders the word as "servants" rather than slaves. When discussing the work of others, I use their translation of the term, but in my own analysis I choose to translate *douloi* as "slaves." The translation of *douloi* as "servants" in a contemporary context evokes images of household servants who are willingly employed by members of a household and paid wages. This kind of economic relationship did not exist in antiquity. In a fun-

damental sense, what distinguished slaves in antiquity from free persons was that they were answerable with their bodies to their masters, subject to corporal punishment as well as sexual abuse. Thus, the translation of *douloi* as "slaves" is important as a signal of these implications. For more on this, see Jennifer A. Glancy, *Slavery in Early Christianity* (Minneapolis: Fortress Press, 2006).

29. Origen, *The Commentary of Origen on the Gospel of St. Matthew*, trans. Ronald E. Heine (Oxford: Oxford University Press, 2018), 658.
30. Origen, *Commentary*, 660.
31. Origen, *Commentary*, 663.
32. Origen, *Commentary*, 664–65.
33. Origen, *Commentary*, 664.
34. Jacques Le Goff, *Your Money or Your Life: Economy and Religion in the Middle Ages* (New York: Zone Books, 1988), 36.
35. Le Goff, *Your Money or Your Life*, 26.
36. Le Goff, *Your Money or Your Life*, 25–27.
37. Le Goff, *Your Money or Your Life*, 27.
38. Thomas Aquinas, *Commentary on the Gospel of St. Matthew*, trans. Paul M. Kimball (Camillus, N.Y.: Dolorosa Press, 2012), 813.
39. Aquinas, *Commentary*, 814.
40. Aquinas, *Commentary*, 815.
41. Aquinas, *Commentary*, 826.
42. Benjamin Nelson, *The Idea of Usury: From Tribal Brotherhood to Universal Otherhood* (Princeton: Princeton University Press, 1949), 30.
43. Nelson, *Idea of Usury*, 31.
44. Nelson, *Idea of Usury*, 78.
45. Nelson, *Idea of Usury*, 85.
46. Nelson, *Idea of Usury*, 108.
47. Deuteronomy 23:20–21: "On loans to a foreigner you may charge interest, but on loans to another Israelite you may not charge interest, so that the Lord your God may bless you in all your undertakings in the land that you are about to enter and possess. If you make a vow to the Lord your God, do not postpone fulfilling it; for the Lord your God will surely require it of you, and you would incur guilt."
48. Nelson, *Idea of Usury*, 119.
49. Whether the parable goes back to the historical Jesus or Matthew/the early church is unimportant for this analysis, but in either case, its historical context is the first century.
50. Justin Ukpong, "The Parable of the Talents (Matt 25:14–30): Commendation or Critique of Exploitation?: A Social-Historical and Theological Reading," *Neotestamentica* 46, no. 1 (2012): 197.

51. Arland Hultgren, *The Parables of Jesus: A Commentary* (Grand Rapids, Mich.: Eerdmans, 2000), 275.
52. Hultgren, *Parables of Jesus*, 278. Hultgren reads the parable as an allegory and reasons that "what is given is what the master considers appropriate. Nothing is given that is more than one can manage," (278) and that "wherever God's gift has already borne fruit, God gives in greater abundance; where it has been fruitless, it is lost completely" (277).
53. Hultgren, *Parables of Jesus*, 280.
54. Luise Schottroff goes so far as to say that he "refused to be a henchman in the dispossession of small farmers"; Schottroff, *The Parables of Jesus* (Minneapolis: Fortress Press, 2006), 223.
55. Rohrbaugh, "Peasant Reading," 35.
56. See, for example, Glancy, *Slavery in Early Christianity*, 102–29.
57. Kathryn Tanner, *Christianity and the New Spirit of Capitalism* (New Haven: Yale University Press, 2019), 30.
58. See Tanner, *Christianity and the New Spirit of Capitalism*, 63–85.

❧ Creeps of the Apocalypse: Climate, Capital, Democracy

CATHERINE KELLER

I am not name-calling. The titular creeps are not personalities but temporalities. They suggest how rapidly the material evidence of global warming accumulates—and how slowly it registers publicly, even now rarely flaming into the news cycle. Yet the time of anthropogenic climate change, read as *a* time, an age among ages, a geological epoch—anthropocene, capitalocene, pyrocene, plantationocene, or otherwise obscene—seems to move not at a creep but a gallop (okay, I admit it, horseman of the Apocalypse). So, a tragic paradox has come into play. The climate crisis is now moving too fast for any full-fix reversal; and at least for a wide and not only denialist public, too slowly to grasp. Add to the problem of its speed-creep this perplexity: there is no responsible way of grasping climate crisis without entangling its science, already dangerously complex for popular consumption, in the divergent temporalities of politics and economics. So, the temporality of climate change pulses in an eerie mix of the imperceptibly slow closure of the epochal window and the accelerating accumulation of planetary effects.

Very differently, the time of politics throbs at the rate of daily news and periodic elections. As U.S. democracy lurches spasmodically toward new electoral deadlines, it casts white supremacist shadows of creeping authoritarianism. That political creep does not signify any predetermined outcome but rather what William Connolly calls "aspirational fascism."[1] He does not foresee its victory but great struggles to prevent such. And of course, the fragility of democracy constitutes a now multinational phenomenon. In this menace our political moment shares, indeed intensifies, the high-pressure threat of global warming. Not surprisingly a rhetoric of apocalypse—not necessarily as "the end of time" but as the catastrophic disruption of human history—now edges even sober secular fears for the democratic and the ecological future. So, in the intersections of politics and climate, divergent speed-creeps form a complicated planetary temporality, over-determined and yet indeterminate. Repair is already improbable and still possible.

But what of the third temporal creep, that of economics? In its current form economics funds both the ecological and the political threats yet presents itself as unthreatened. Far from the disruptive discontinuities of ecopolitical time, it expresses an opposite sort of temporality: that of the smooth continuum of assured progress. The current global capitalism depends upon one party's war against corporate regulations and its white evangelical climate denialism. And at the same time, it promises that "Capitalism is the Key to Fixing Climate Change."[2] This ever more green-washed capitalism creeps from a determining past into its guaranteed future by way of a resiliently continuous present. Indeed, in Kathryn Tanner's theological response to "the new spirit of capitalism," it is this temporality, with its self-reproducing continuum, that defines the economic dynamism of the current neoliberalism.

In other words, the *ecological* repair that is still possible demands improbably immense and fast *political* change to slow—quickly—the *economic* drivers of the crisis. Does this triple time not appear to be materializing as a speed-creeping apocalypse? I am here imagining that those three asymmetrical temporalities might be helpfully read in their systemic entanglement—from a mindfully multitemporal and at-the-same-time perspective. Such a polyrhythmic point of view would entail in this case the thought experiment of a theology not just multidisciplinary in perspective but *trans*disciplinary, aiming through and beyond disciplinarity itself toward practice—indeed, toward practice *in time*— quickly enough to make a difference but slowly enough to avoid paralyzing panic. Such theology takes place not outside of time in a heavenly retreat center, nor in an end-of-time headquarters.

If in this case practice in time demands a bit of theological attention to the ancient future temporality of John's *Apocalypse*, it also attends to the present precarity of theology itself. It knows that its own *trans* could swing all too apocalyptically into the creeping death of *theos* and therewith of theology.³ Alternatively, that theo-precarity might release the uncertainty, the opening, of a more plural, volatile possibility. In its contracted multi-temporality, the tense present appears to crack open a space of planetary indeterminacy, a space (creepily) framed by apocalyptic *over*determination. Speaking theologically: *apokalypsis* read responsibly, even correctly, in context, does not close. It "unveils." It dis/closes.⁴ The exercise of this meditation therefore folds the temporalities of climate, beset by capital and betrayed by democracy, into the transdisciplinary opening of a polyrhythmic present. And because such disclosure is steeped in an ultimate concern that some of us still nickname God, the present tense keeps opening—however tensely—into possibility. No matter what.

ECONOMIC ENDGAME

"A kind of unbreakable continuity exists between past, present and future—they in fact collapse into one another, fuse with one another—in ways that make any radical break with the present order seem impossible." Thus, Tanner characterizes the current capitalist temporality. The financialized transactions of the present economy operate at split-second virtual speeds—and yet move as though unchangeable. "This is a constantly changing economic order or regime, requiring constant change from the participants in it, but one offering no escape from it; the future simply promises more of the same." So, in fact "the present monopolizes attention in ways that chain one to it."⁵ We might call it the "propertied present." It is founded on a classical substance metaphysics, where the subject—as first, man of substance, sovereign over his properties—presents at the creep-speed of an essentialized self-identity through time. Reproduced in endless innovations, risks, and resiliencies, the smooth time of capital can capitalize every difference by remaining commodifiable as the same. Endlessly. And so even in view of the meltings, the floodings, the burnings: it has no end in sight.

Thus, as a recent secretary of state, Mike Pompeo, announced in remarks in Finland, melting sea ice presents "new opportunities for trade." He continued, "The Arctic is at the forefront of opportunity and abundance. . . . It houses 13 percent of the world's undiscovered oil, 30 percent of its undiscovered gas, an abundance of uranium, rare earth minerals, gold, diamonds, and millions of square miles of untapped resources, fisheries galore."⁶ This winning optimism thus monetizes melting glaciers.

In the face of planetary volatilities of climate, politics, a virus, a war, and despite a capital-driven economic melt-down as recent as 2007, the current form of financialized capitalism projects the image of a propertied present continuously progressing. Therefore, its growth can only be conceived as infinite. The very notion of some limit to the market's growth, let alone some ecological or democratic constraint, remains economic heresy.

Theology, in several minor keys, has labored to expose the secularized infinite as the all-propertied idol of capitalist orthodoxy—and at the same time the creeping Christian collusions with the political and economic status quo. Over half a century ago there arose in the global South the great movement of Roman Catholic liberation theologies on behalf of the poor and eventually also, with Leonardo Boff and Ivone Gebara, of the Earth; while in the North, Jürgen Moltmann and John Cobb early pitted Protestant theology against the jointly ecological and human destructiveness of the capitalist absolute.

As Joerg Rieger nails the problem, "We now understand that where the goal of production is the generation of profits—the categorical imperative of the Capitalocene—production must drive consumption, evolving in a vicious cycle that results in ecological destruction and climate change."[7] So global warming and the global means of production cycle dangerously around the planet. In *Of Divine Economy: Refinancing Redemption*, Marion Grau reconstructs the economy theologically for the sake of an ironic *commercere* as "common investment," seeking to undo the self-centered and domesticated divinity of late capitalism.[8] Exposing its increasingly dominant financialized form in *Christianity and the New Spirit of Capitalism*, Kathryn Tanner reads with and against Max Weber to show how this spirit feeds on a dominant form of puritanical Calvinism.[9] She pursues her critique not as a sociologist but as a theologian, in precise fidelity to her "own, quite specific Christian commitments."[10] These diverse theological critics all discern the spirit of an alternative *oikonomia*, a divine economy vibrating with the ancient possibility of an actually, radically, *common* good. That polyrhythmic possibility has not lacked in revolutionary and evolutionary actualizations—even amidst that good's buyouts and sell-outs on behalf of the goodies of the competing economic absolute.

In the meantime, the propertied present continues to accept worship, religious and secular, whereby it is granted its due: faith in its creeping infinity of growth. Funding the household, the *oikonomia*, of its economic elites, its propertied present despoils in this time the space, the *oikologia*, of the common habitat of its species. So, the present economy assembles at creep-speed the new figuration of an "uninhabitable planet."[11] Because the *eco*logy of a hospitable earth can no longer coexist with the present form of the *eco*nomy, the self-contradiction within the *oikos* clashes against the temporality of capitalist gradualism into the accelerating volatility of the planetary habitat. Capitalist time has conquered earth's space. It is not that time itself dominates space. Rather, the infinitude of the propertied present tense of a few jeopardizes the habitable space of a majority. And the finitude of spatiotemporality manifests ever more desperately—at least as concerns the habitat hospitable to most humans and nonhumans.

As Bruno Latour notes, the economic elites now know this. They no longer pretend that wealth will trickle down to all. They have "understood that, if they wanted to survive in comfort, *they had to stop pretending, even in their dreams, to share the earth with the rest of the world.*"[12] They have realized there isn't enough space, enough Earth, enough stuff, for us all; there will be less and less, for more and more people. They have begun to prepare their own climate change households in lovely white locations up north, thus offsetting any looming guilt as to their grandchildren's futures even as they make their high-speed profits at the future's expense. In "How Big Business Is Hedging against the Apocalypse," Jesse Barron warns (unsurprisingly) that the hedging is not in ways that environmentalists will like. Without even the excuse of ignorance, major companies like ExxonMobil continue to profit from climate destruction.[13]

Does the future thus capitalized rule out the high-speed temporality needed for a political turnaround? Toward—what shall we call it? Ecosocial justice, a slowing of climate change, a revolution in economico-political practice? The salvation of a whole bunch of species, including even ours? The salvaging of the yet habitable planet? At least some earthwide assemblage of and for a *common habitability*? And yet. Any such turnaround would require, as two major IPCC reports made grimly clear, international cooperation for an economic withdrawal from fossil fuels requiring political transformation larger in scale and faster in speed than anything known in history.[14]

How probable is that? Inasmuch as chances can be calculated and the future predicted, the Game Over chorale will only swell in volume. Under these eerie circumstances, the new normalization of a perfectly secular vocabulary of ecological apocalypse—of glaciers melting, coastlines flooding, fires around the globe raging—can only spread. I used to disdain most public uses of "apocalypse" as extremism. But then it seemed to me that this public eco-apocalyptic vocabulary broke beyond any tones of hysteria with the headlined studies called "Insect Armageddon" and then "Insect Apocalypse" of the two Americas. The insects form the basis of our food chain. Their die-out portends our own.[15] If only because of the ancient imaginary of planetary catastrophe, theology creeps, not quite visibly, into a new public speech of the unspeakable. *Apokalypsis* blurs into *apophasis*: an apophatic anthropocene that no mystification can save from the unspeakably destructive impact of its *anthropos*.

The relation between the volatile ecological speed-creep and the relentlessly creeping capitalist infinite has locked into the internal contradiction of the Anthropocene itself. The power of a small portion of the *anthropos* over the Earth increasingly turns Earth against the *Anthropos* at large. Planetary systems unfortunately cannot aim the effects of man-made change at the exceptional wealthy white man. And that economic exceptionalism invests in political support of, for instance, the white U.S. (and Christian and heteromale and successful) exceptionalism. As political theology has demonstrated already at its Schmittian root, it is by manipulating emergency that the exception makes itself the rule.[16] So, power can then take advantage of the emergencies—of climate, of class—that it causes. Sovereign power democratically elected may turn autocratic, indeed fascist, precisely in its nationalist manipulation of a white supremacist and largely non-elite public.[17]

As to that whiteness, another important, underacknowledged correlation between politics and economy unfolds. It happens between whiteness and class, and it happens *ecologically*, in what J. Kameron Carter all too accurately names the "plantationocene." Carter summons for a time of climate change what W. E. B. DuBois called "the propertization of the world."[18] Already a century ago DuBois captured its political theology, indeed the ecologico-economico-political theology, with prophetic sarcasm: "Whiteness is the ownership of the earth forever and ever. Amen!"[19] From the present angle of time a polyrhythmic theology cannot co-opt but may echo Carter's articulation of the needed turnaround: to a "'non-class culture' that heralds nothing less than the apocalypse of the (plantation as) World." No straight "end of the world," this apocalypse—but the doom of a civilization that has rendered the earth its plantation.

Carter's stunning "Black Malpractice" can be read as a new manifestation of a long history of revolutions. As the Marxist philosopher Ernst Bloch has demonstrated in *The Principle of Hope*, that history translates the apocalyptic radicality of the biblical legacy into political motivation—into a collective refusal of the organization of culture by class. To face that civilizational economy as it antedates and produces whiteness and blackness as well as capitalism, the "principle of hope" depends now upon a persistent confrontation of the entanglement of race and class.

White world without end: is that the face of the propertied present and its infinite growth? No matter how many ethnic diversities and skin tones it can commodify, advertise, and liberally monetize? The racial register of the propertied present that came into its own through the Euro-American slave trade brands the half millennium old beginning of the plantationocene as Modernity itself. Thus, as Tanner observes, it partakes theologically of the early modern temporality that "in a traditional Protestant ethic also finds an insidious analogue in the new spirit of capitalism."[20] Not limited to the Puritan experiment but ecumenical in its Christian capture of the Americas, this half-millennium–old modernity hardly pauses in its self-replication at any portal to the postmodern. So, it speed-creeps along as though endless, self-continuously racing, raced, through various ends of the modern: differences, dispossessions, and disasters. All cashed out and reinvested in its propertied present, its proper infinity.

PORNE-TRADE

Now, however, we note a recent, still undertheorized shift in the relation of economics to politics. We must not miss the deep tension between, on the one hand, global capitalism in the neoliberal format of the past half century, the form that we on the academic left presume and criticize, and on the other, the white nationalist—*anti-globalist*—politics with which that capitalism converges in recent U.S. history.

Here an apocalyptic anachronism creeps up on me. One of the most garish parodies presented in the book of Revelation stages what can be read as an ancient form of the tension of politics and economics: the Beast of empire riding "the Whore (*porne*) of Babylon." The beast symbolizes the Roman *polis*. The "Great Whore" embodies another dimension of empire: global trade. The text cashes out in detail the lament of the urban and marine merchants at her fall, and at the loss of *twenty-six* luxury products of the ancient world: "cargo of gold, silver, jewels and pearls . . . purple dye . . . wine, olive oil . . ." with "slaves—human lives" providing the climax of the list (Rv 18:11–13).[21] There could hardly be a clearer symbolization of a global economy.

Of course, the image, like the text as a whole and like most of its world, colonizer or colonized, partakes of a systemic misogyny. When toward the end of the last millennium I wrote *Apocalypse Now & Then: A Feminist Approach to the End of the World*, the vilification of that female figure as slut distracted me from its economic radicality. (I befriended her, called her Babs, blamed John for blaming economic imperialism on her.) But in my recent *Facing Apocalypse*, I recognize that the author's aim is not to strengthen male supremacy as such—he hardly needs to. With the flashing image of the imperial prostitute (given her sovereign status one can hardly call her a "sex worker"), the Great Babylon vision cultivates horror at the limitless commodification—unto the life of one's own most intimate flesh—imposed by the sovereign globalization of trade. And as today, that commodification imposes itself not by extrinsic force but by the seductive formation of desire. It is in part because of its indubitably creepy sexism that the prophetic force of Revelation's exposure of the pre-capitalist form of imperial global economy as idolatrous has been underappreciated by the Christian left.

When the beast then turns on his porn queen and devours her, is John not revealing a deep contradiction between the political and economic dimensions of a single global regime? Thereafter the text's nightmare spiral through the destruction of immense systems of human and nonhuman life accelerates—fire consumes "one third of the trees," poison takes out "one third of the life of the seas" (8:7f). . . . Of course, the vision does not literally foresee climate change, nor yet the specific neoliberal capitalist form of the mating of economics and politics. Prophecy is not prediction. It does not foretell a somehow predetermined future. But it does read unjust patterns already so deeply entrenched in its present as to make their systemic iterations in unknowable futures all too probable.

How Will Capitalism End? by German sociologist Wolfgang Streeck helps to update the contradiction. He finds current Western democracies undergoing a risky struggle between two constituencies not simply aligned with right and left: the "national state people . . . and the international market people."[22] The tension comes to a head because capitalism "can only survive by constraint of its own complete commodification" of land, labor, and money. Yet it is motivated by utterly unconstrained, insatiable "growth." And there is no regulatory motivation or structure on the scene with the power to insist upon limitation.[23] As the contradiction took its late twentieth-century course, "the postwar shotgun marriage between capitalism and democracy came to an end."[24] Does that marriage and its collapse not faintly echo the whore-beast parable?

Yet it would be misleading to infer that the collusion of global capital with national politics has come close to ending in the twenty-first century. Rather, the take-down of democratic, eco-socially oriented regulation of corporations is itself key to conservative politics. That is no take-down of the global economy! So, in fact the right wing of U.S. politics depends upon the neoliberal globalism that it at the same time attacks. The authors of the fundamentalist Christian *Trumpocalypse* (a term of praise) align "global capitalism" with all cosmopolitan efforts: such as the UN, interchange between world religions, open borders, and of course the ecological movement. So, Trump, they declare in apocalyptic celebration, "is one of the few politicians who is at war with globalism. Most of the Republicans and Democrats, along with Obama and the Clintons, for all intents and purposes, are on the payroll of the wealthy corporate elite."[25] The latter claim may have some credibility; but that Trump had any political power apart from his extensive global business partnerships, alliances, and priorities has no basis in fact.[26] And of course, the role of conservative white evangelicals in keeping political Beast and economic Babylon coupled remains indispensable.

So the nationalist/globalist partnership is not over in the U.S. but rather has morphed, in the political philosopher William Connolly's language, into a particularly "virulent evangelical/neoliberal assemblage." He analyzes how "white evangelicals foment an accusatory spirituality toward nonwhites, ecologists, the media, migration, independent women, decolonial regimes, and intellectuals." In this they construe the market as "a vehicle of God's providence" while they hold "the democratic state to be a bureaucratic titan."[27] So, the tensive intimacy of Babylon and Beast keeps twisting into self-contradiction. The right-wing assemblage praises the present form of capitalism, even as it derides the globality of neoliberalism that *is* that form, and when it can attacks it by means of the titan.

DISRUPTIVE NOVUM

The complication, or co-implication, of democracy and capitalism only intensifies the spiral of ecosocial demise. The contradictions between the volatile times of political polarization and of climate shifts, trapped in the continuous capitalist present that monetizes both, press toward greater emergency. As political theology makes clear, emergency justifies exceptionalism. So, with emergencies racial, national, economic, the vicious spirals spin faster. The very spin holds the exceptional subject of the propertied present continuous—indeed sovereign. Times of crisis may simply reinforce the creep of the current form of financialized capitalism.

As Kathryn Tanner persistently argues, "The more people are convinced of derivatives' capacities to tame the future, the more imprudent they become, taking more risks than they should, so that it becomes all the more likely that the future will bring catastrophic surprises." And "that unanticipated catastrophic surprise closes off all possible options."[28] Note that such closure would characterize not the darkly *dis*closive apocalypse of J. Kameron Carter (or of John of Patmos) but the banal *fin* of a deluded infinity. In other words, the steady creep of the propertied present, confident in endless growth, conceals the recklessness of its gambles. To read the catastrophic potential of the propertied present as intersected in its illusory self-sameness by the speed-creeping temporalities of climate change and political polarization only underscores Tanner's economic analysis.

That catastrophic potential comes here edged with theology. Indeed, it is just at the edge, the *eschaton*, of Tanner's theological vision that another register of catastrophe is revealed. The vision exposes not just the reckless financialization but the religious delusions of mere rescue from time and its volatile indeterminacies. Here "the future that Christians expect retains a strongly negative flavor of potential disruption, something whose effects it would therefore make sense to try to nullify prospectively, and not just because Christians sometimes think that future might include damnation for some. Even a purely benevolent end—universal salvation—retains a highly negative cast to the extent such a future will tear us away from what we remain in and of ourselves, sinners."[29] In resistance to such redemptive dissociation from present materiality, Tanner's own soteriology comes into play precisely at that edge of time—whatever ends of whatever worlds suggest themselves. "Salvation would not simply await the resurrection of bodies to come but would be operative now to transform material lives for the better."³⁰

Of course, even a theology or a spirituality of material transformation can serve capitalism. One passing example bubbles up: in Kathryn Lofton's narration of the merger of religion and consumption in the marketing of soap a century ago, "soap offered not only sanitation but explicit salvation." Pushing the Protestant merger of cleanliness and godliness, a reporter comically raved, "There are millions without the desire to repent and be *laved*. We must continue to 'sell' the world on cleanliness."³¹ Alien to such sudsy consumer positivism, salvation in its biblical sense never lacks a critical, "strongly negative flavor." It does not abstract life from the fray nor persons from the bodies and the dirt of our shared life. No new heaven and earth result from individual purity or manipulable transition.

Tanner offers instead the "coherence of a whole new world to be entertained as an imaginative counter to the whole world of capitalism as it presently exists and pretends to be all-encompassing, to have no limits, nothing outside itself." Such a possible world operates not at a remove from finance-dominated capitalism but by cutting across it, traversing it to disruptive effect."³² With and without an explicit theos, kataphatic or apophatic, such a disruptive novum offers a possible clue to a hope against hope, against the falsities of optimism, progress, guarantee.

Such an embrace of the negative, not as a manipulable dialectic but as disruption, resonates with Carter's "black rupture." It is "at the limit of politicality" that his edgy *eschaton* cuts across the present. There "the surreal—the subreal, the submerged—flutters as invisibly felt and is apocalyptically unveiled as monstrous, as a sacral blackness that incites volatility." As rupture opens into black rapture, the propertied present pales and shakes. Of course, this *apokalypsis* of race dynamics does not unveil a generic Christian "new world." But neither will it dissociate from the counter-economics, counter-politics, counter-theology of ancient prophetic eschatology.

Apocalyptic unveiling in our time and in U.S. space doesn't happen outside the spatiotemporality of a racialized modernity in which I—for one—forfeit all innocence. So of course, one may creep off into a silence that no apophasis excuses. Before an *apokalypsis* that no revision assuages. Rupture never guarantees rapture. And yet economically and ecologically, *oikos* imagined otherwise might dis/close, in J. Kameron Carter's words, "a zone of festive dwelling." Such a dwelling is hardly dissociable from the dream of urban newness that concludes of the old Apocalypse. With its cross-species eroticism (bride/lamb—in ironic juxtaposition to whore/beast), the vision of New Jerusalem celebrates a spacetime of dwelling, all twelve of its gates festively open 24/7. But in the meantime, this mean time, great gates of possibility seem to be creaking toward closure.

SALVAGING APOCALYPSE?

Might we agree that the triple temporality of the present—ecological, political, economic—does not assemble any merely predictable future? Even though as you read the tense present of an unpredictable degree of ecosocial catastrophe all too predictably kicks in? The ineradicable indeterminacy of any moment blurs with an unspeakability that at its theological depth means not horror but mystery. Apokalypsis and apophasis, so opposite in mood, won't here break apart. Even so, the unspeakable horror of the apocalypse swerves into its over-imagined and under-realized hope for the future, no less difficult to name. That promising future hides in an apocalyptic apophasis. So then the unclosed margin of indeterminacy does mean at least this: catastrophe can become catalyst. Climate crisis with its socioeconomic consequences carries a potential for disruption that movements of resistance on their own can hardly match. Yet we can have no certainty that heightened ecological emergency will somehow provoke the emergence of (for example) a dark green democratic socialism. The dangers do not stand down: not just in the refusal of responsibility by those reaping profits from business as usual, but in some version of climate fascism claiming emergency power. And then there are the risks posed by the last-ditch mitigations and big technofixes. And when the fixes fail, with possibly horrific side-effects, we face the dangers of humanly—nationally, racially, sexually, tribally—brutal adaptations. On a burned, baked, and flooded planet with some gated communities for the self-designated exceptions. "New Jeru" for the few.

So, what possible hope (oh, that discredited notion) might keep us not doped but creatively struggling? What glimpses of possibility will press us into polyrhythmic solidarities, outpacing the triply creeping dooms of economics, ecology, democracy? Surely such hope carries a soteriological intensity. That does not mean it comes down to dependence upon intervention and salvation by an omnipotent Other, let alone to a final climax.

Salvation might redistribute its agency more democratically, its matter more ecosocially, its sacrality more relationally. This does not mean, within this theological *oikos*, atheism. Salvation does, however, etymologically ally itself with "salvage."

Salvage happens to be the name of a British "quarterly of revolutionary arts and letters." A recent issue sports as subtitle *"Towards the Proletarocene."*[33] The journal's "salvage communism," characterized in the words of novelist (and journal co-editor) China Miéville as an "apophatic Marxism" freed of all Stalinizing dogmatism, holds in political conjunction the depredations of capitalism, racism, and, "as we race past tipping point after tipping point," climate change. "That is to say, global proletarianisation and ecological disaster have been products of the same process. The earth the wretched would—will—inherit, will be in need of an assiduous programme of restoration."[34] One need not convert to Marxism to take in Miéville's disruptive poetics, as he meditates in Suffolk— "its lines of dead trees, lichened concrete in the bird sanctuary, drowned houses off the coast"—on "a nationally sanctioned becoming-eerie" of "a catastrophe already here." Hope, in other words, cannot be to avoid catastrophe. "To hope against hope is not merely to contest all the dreg-like hope unearned: it is prefiguration. It is to live with an eye on the horizon where hope will be no longer needed."[35] And in the light of that prefiguration to salvage, to discerningly recycle, that which can still be saved.

Might hope thus fold its futures into the *oikos* of festive dwelling? A New Jerusalem inviting no religious or political certitudes? Prefiguration does not mean that we are already progressing toward it: "Interruption could hardly be more urgent."[36] Rupture opens possibility now and does not wait to reach any horizon. After all, a horizon only exists by receding. The prefiguration transfigures the present, which, dispossessed of its properties, opens to the temporalities of its tenses: a present contracting in itself tragicomic remembrances of what has been, even while imagining what is yet to come and even now becoming. Its present does not escape the tenses of past and future. Indeed, those tensions of time itself are intensified by the creeps of the apocalypse. The contradictions between the possible futures churning up out of the conflictual past will continue to threaten to overwhelm the present. And yet rather than surrender to impossibility or paralysis, we might salvage even the contradictions, the disappointments, the hopelessness—and transmute them moment by moment into a mobilizing complexity of time-tempos. There the Marxist utopia and the biblical New Jerusalem do a circle dance together.

PLANETARY KAIROS

In this polyrhythmic now, possibility foments the new—not the old new of endless consumption, selfsame in its dissociative difference, but an earthy freshness that materializes now and here. Salvaging is not a matter of clinging to the ruins of the past. The operative prefiguration is neither a refiguration of the given nor a prediction of the future. Its novum arises not as exception but as *inception*.[37] In other words, we might learn to salvage from the past whatever—however impurely—can be recycled renewingly. In the polyrhythms of labor, lament, love. And for the foreseeable future—of hope.

Can the elemental convulsions of planetary time now open up (*apokalypsis*) another temporality, far from The End of Time, tuned to the volatility of the current climate, material and political? Might its assemblages salvage from their determinate history the possibility of habitable earth futures—*in time*? Hope for that possibility lives free of the progress optimism of capitalism. It may remain difficult to distinguish from ecopolitical pessimism. But the difference matters, it materializes in this salvaging that is the very work of ecology, vastly prehuman, more than human, and imperatively now-human. It lets us imagine—in rupture, in rapture—timely works of collective creativity. And of such creative process, let me stress all too theologically: creation is not ever from mere nothing.[38] Some dark energy, some pulsation of a pneumatological wing over the rhythm of ocean, is always already in play. Inception comes not as an omnipotent performance of an ex-nihilo, not even secularized as revolution. It revolves, recycles the potentiality of the past as present potency, thus strengthening the *radix*, the rooted earth-time, of honestly radical revolution. In that ground it cuts off no ancient source of justice, peace, and new creation. For its turbulent depth supports the needed breadth. That breadth becomes practical in the assemblage of a wide enough public to salvage the planetary habitat. Transformation then carries glimmers and graces of old prophetic soteriology—salvation not as supernatural flight but world-salvaging. The prophets meant to activate and to liberate, not to discipline and postpone.

If we salvage some theological sense of salvation itself, the haunting hope of new creation finds spiritual reinforcement. It finds spirit—with which to call up secular-religious assemblages, lively in their intersections.

As Marcia Pally demonstrates, for a new economics and politics, "we need a way back to the evolutionary, ontological and theological principle of distinction-amid-relation."[39] Solidarity does not need us to come together in some solidified unity, ontological or social. It may require—and stir and suffer—immense differences of priority. It does not wait for a unification of the desired versions of political equality and economic justice and ecological restoration but goes ahead and practices a pluralizing entanglement of our histories and our futures. Amidst the triply creeping temporality, solidarity arises not only in the simple "against," holding out for some miracle of redemptive inversion. Sometimes it can only salvage scraps of potentiality from inadequate gestures, well-intended compromises, unpracticed theories, broken promises. It starts always now, again, from the chaos, the detritus, in the improvised polyrhythm of our spirited planetarity. In this spirit it can crack open the actual opening within the speed-creeping apocalypse of ecology, economics, politics.

At the same time, the transdisciplinary *oikos* of such theology continues, along with much liberal education, its own not so slowly creeping demise. Such fragile institutions tell their own lamentable, laborious, and occasionally loveable stories. And still the speed of our theory and the creep of our practice must keep pulsating with the questions of climate, capital, and democracy. Is every such theological effort doomed to reflexive curation of the apocalypse? Perhaps. And so we must keep reinforcing the inceptions, the dis/closures, that put the lie to The End.

The disruptive moment of this planetary *kairos* of last chances will enable a widening solidarity, breaking out of the marketable modules of faith and knowledge. And at this time, this time that stretches from this writer to this reader, the contradictions of economics, politics, and ecology may yield to an earth-wide temporality. Neither creeping nor galloping, this moment depends upon a spatiotemporal indeterminacy ripping open perilously new possibilities. In that volatile dis/closure, improbable futures still stir, yet to be salvaged in the inception of their polyrhythmic present.

NOTES

These ideas are developed further in *Facing Apocalypse: Climate, Democracy and Other Last Chances* (New York: Orbis Books, 2021).

1. William Connolly, *Aspirational Fascism: The Struggle for Multifaceted Democracy under Trumpism* (Minneapolis: University of Minnesota Press, 2017).

2. "A Rebuttal: Ronald Bailey, "Capitalism Is the Key to Fixing Climate Change," *Reason: Free Minds and Free Markets* (September 9, 2019). https://reason.com/2019/09/20/capitalism-is-the-key-to-fixing-climate-change/printer/.
3. Another register of "trans" unfolded at the student-led Transdisciplinary Theological Colloquium at Drew 2018: *Trans: Human/Divine Bodies beyond Boundaries*, http://2018.drewttc.com/.
4. For an in-depth exploration of "apocalypse" as dis-closure, see Catherine Keller, *Facing Apocalypse: Climate, Democracy, and Other Last Chances*.
5. Kathryn Tanner, *Christianity and the New Spirit of Capitalism* (New Haven: Yale University Press, 2019), 28.
6. Mike Pompeo, who adds, "Indeed, we have quietly entered a dangerous competition with Russia for Arctic resources"; "Jennifer Hansler, "Pompeo: Melting Sea Ice Presents 'New Opportunities for Trade,'" *CNN*, May 7, 2019—too slow to compete for most news cycles, except now and then when the fires of California, Brazil, or Australia blaze briefly into the headlines, let alone to compete with the time-beating sensationalism of commercials (politics generates and capitalism manipulates a far more dramatic temporality); https://www.theguardian.com/environment/2020/jan/09/white-house-projects-permits-climate-impact-plan https://www.alternet.org/2020/01/winter-isnt-coming-and-you-need-to-prepare-for-the-pyrocene/?utm_source=&utm_medium=email&utm_campaign=3443.
7. Joerg Rieger, *Theology in the Capitalocene: Ecology, Identity, Class, and Solidarity* (Minneapolos: Fortress Press, 2022), 40f. See also my response in Vanderbilt Divinity School Newsletter of the Wendland-Cook Program in Social Justice, September 2022, https://www.religionandjustice.org/interventions-forum-theology-capitalocene.
8. Marion Grau, *Of Divine Economy: Refinancing Redemption* (London and New York: T & T Clark/Continuum, 2004).
9. Tanner, *Christianity and the New Spirit of Capitalism*, 1–7.
10. Tanner, *Christianity and the New Spirit of Capitalism*, 7.
11. David Wallace-Wells, *The Uninhabitable Earth: Life after Warming* (New York: Delacorte Press, 2023).
12. Bruno Latour, *Down to Earth: Politics in the New Climatic Regime* (Cambridge and Medford, Mass.: Polity Press, 2018), 19.
13. Jesse Barron, "How Big Business Is Hedging Against the Apocalypse," *New York Times Magazine*, April 11, 2019.
14. The IPCC gives us about a decade from now. See my reflection (with Agamben's) on Paul's "time that remains," in Catherine Keller, *Political Theology of the Earth: Our Planetary Emergency and the Struggle for a New Republic* (New York:

Columbia University Press, 2018), 51–55. Read the IPCC reports here: https://www.ipcc.ch/reports/.
15. See Rachel Ramirez, "Parts of the World Are Headed toward an Insect Apocalypse, Study Suggests," *CNN*, https://www.cnn.com/2022/04/20/world/insect-collapse-climate-change-scn/index.html and https://www.nytimes.com/2018/11/27/magazine/insect-apocalypse.html.
16. "Sovereign is he who decides on the exception"; Carl Schmitt, *Political Theology: Four Chapters on the Concept of Sovereignty*, trans. George Schwab (1985; Chicago: University of Chicago Press, 2005), 5.
17. See Keller, *Political Theology of the Earth*.
18. J. Kameron Carter, "Black Malpractice," *Social Text* 37, no. 2 (2019): 67–107.
19. W. E. B. DuBois, *Souls of Black Folk* (1903; New York: Dover, 1994), 16.
20. Tanner, *Christianity and the New Spirit of Capitalism*, 28.
21. Keller, *Facing Apocalypse*, 111–14.
22. Wolfgang Streek, *How Will Capitalism End? Essays on a Failing System* (New York: Verso, 2016), 24.
23. Streek, *How Will Capitalism End?*, 16. Streek writes, "By attending to the need for democratic political legitimacy and social peace, trying to live up to citizen expectations of readily increasing economic prosperity and social stability, [government policies] found themselves at risk of damaging economic performance." At the same time, efforts to prioritize economic growth tended "to trigger political dissatisfaction and undermine support for the government of the day and the capitalist market economy in general."
24. Streek, *How Will Capitalism End?*, 20: "That marriage was made not in heaven but in the urgency of establishing a new order after the catastrophe of two world wars (make deals not war) and in the face of the communist competition."
25. Paul McGuire and Troy Anderson, *Trumpocalypse: The End-Times President, A Battle against the Globalist Elite, and the Countdown to Armageddon* (New York: FaithWords, 2018), 97f: "They sold America down the river long ago—as did their EU counterparts—with numerous trade treaties that promote globalism."
26. "Trump's Global Web of Partners," *Forbes*, https://www.forbes.com/trump-global-partners/#689851476859.
27. William Connolly, *Climate Machines, Fascist Drives and Truth* (Durham, N.C.: Duke University Press, 2019), 54f.
28. Tanner, *Christianity and the New Spirit of Capitalism*, 156.
29. Tanner, *Christianity and the New Spirit of Capitalism*, 157.
30. Tanner, *Christianity and the New Spirit of Capitalism*, 201.

31. Kathryn Lofton, *Consuming Religion* (Chicago: University of Chicago Press, 2017), 93; italics mine.
32. Lofton, *Consuming Religion*, 219. Tanner adds, "along the very line of the ethics of self-transformation that is the relay or transfer point of its various dimensions, the hinge or axis around which the whole turns, that aspect upon which this entire old world has riveted itself."
33. *Salvage: Towards the Proletarocene* 7 (October 2019).
34. China Miéville, "Silence in the Debris, Towards an Apophatic Marxism," *Salvage* 6 (November 2018): 115–44. Miéville is a great novelist, Marxist historian, and founder/co-editor of this journal.
35. Miéville, "Silence in the Debris," 115–44.
36. Miéville, "Silence in the Debris," 115–44.
37. Catherine Keller, *Political Theology of the Earth: Our Planetary Emergency and the Struggle for a New Public* (New York: Columbia University Press, 2018).
38. See Catherine Keller, *Face of the Deep: A Theology of Becoming* (London and New York: Routledge, 2003).
39. Marcia Pally, *Commomnwealth and Covenant: Economics, Politics and Theologis of Relationality* (Grand Rapids, Mich.: Eerdmans, 2016), 351.

LIST OF CONTRIBUTORS

GARY DORRIEN is an American social ethicist, philosopher, and theologian. He is the Reinhold Niebuhr Professor of Social Ethics at Union Theological Seminary in the City of New York and Professor of Religion at Columbia University, both in New York City. He has authored twenty-one books and more than 300 articles that range across the fields of social ethics, philosophy, theology, political economics, social and political theory, religious history, cultural criticism, and intellectual history.

PAULINA OCHOA ESPEJO is Associate Professor of Political Science at Haverford College. She works at the intersection of democratic theory and the history of political thought, and she is interested in questions about popular sovereignty and borders. She is the author of *The Time of Popular Sovereignty: Process and the Democratic State* and co-editor of *The Oxford Handbook of Populism*.

MARION GRAU is Professor of Systematic Theology, Ecumenism, and Missiology at MF Norwegian School of Theology, Religion, and Society in Oslo, Norway. Her teaching interests are in constructive theology and critical intersectional theories. She is the author of several books, including *Pilgrimage, Landscape, and Identity: Reconstructing Sacred Geographies in Norway* and *Of Divine Economy: Refinancing Redemption*.

EUNCHUL JUNG is a PhD candidate in the Theological and Philosophical Study of Religion area at Drew University. His research interests also include free will, subjectivity, German idealism, psychoanalysis, critique of civilization, and phenomenology.

CATHERINE KELLER is the George T. Cobb Professor of Constructive Theology in the Theological School and Graduate Division of Religion of Drew University. She practices theology as a relation between ancient hints of ultimacy and current matters of urgency. She is the author of numerous books, including most recently *Facing Apocalypse: Climate, Democracy, and Other Last Chances*.

HILARY MCKANE is Director of Graduate Academic Services and holds a PhD in Bible and Cultures from Drew University. Her research interests include economics and the New Testament, parables, and women in the Greco-Roman world.

MARCIA PALLY teaches at New York University and held the Mercator Professorship in the Theology Faculty at Humboldt University-Berlin, Germany, where she remains an annual guest professor. She is the author of several books, including *White Evangelicals and Right-Wing Populism: How Did We Get Here?*, *From this Broken Hill I Sing to You: God, Sex, and Politics in the Work of Leonard Cohen*, and *Commonwealth and Covenant: Economics, Politics, and Theologies of Relationality*.

JENNIFER QUIGLEY is Assistant Professor of New Testament at Candler School of Theology at Emory University. Her research lies at the intersections of theology and economics in New Testament and early Christian texts. She has interests in archaeology and material culture, and her research and teaching are influenced by feminist and materialist approaches to the study of religion. She is the author of *Divine Accounting: Theo-Economics in Early Christianity*.

JOERG RIEGER is Distinguished Professor of Theology, Cal Turner Chancellor's Chair in Wesleyan Studies, and the Director of the Wendland-Cook Program in Religion and Justice at Vanderbilt University Divinity School; he teaches in the Graduate Department of Religion at Vanderbilt University. He is the author and editor of twenty-six books and more than

175 academic articles, including most recently *Theology in the Capitalocene: Ecology, Identity, Class, and Solidarity*.

DANIEL A. SIEDELL is Visiting Researcher and Curator at the Center for Theology, Ecology, and Culture at the Stockholm School of Theology and teaches in the International Master's Programme in Curating Art at Stockholm University. He is a PhD candidate in Theological and Philosophical Study of Religion at Drew University.

DEVIN SINGH is a social theorist, scholar of religion and theology, and leadership strategist and adviser. Devin is a tenured Associate Professor of Religion at Dartmouth College, where he teaches courses in religion, philosophy, ethics, organizational dynamics, and the connections among religion, economics, and politics. He is the author of *Divine Currency: The Theological Power of Money in the West* and *Economy and Modern Christian Thought*.

INDEX

aggression, 171–75
American Federation of Labor (AFL), 27, 29–31, 33–36
amorous agonism, 143, 146–49, 155, 157
antagonism, 142–43, 145–49, 155–58
Anthropocene, 6, 201, 206; anthropocentrism, 15
Antonelli, Paola, 111, 117, 119–22
anxiety, 117, 120, 142–43, 145–46, 152
apocalypse, 4–6, 106–7, 109, 121, 123, 201–2, 205–6, 210, 212–13, 215; eco-apocalypse, 26–27; *apokalypsis*, 203, 206, 211–12, 214
Aquinas, Thomas, 163–64, 166, 184, 192
art, 107–8, 110–118, 120, 122, 125n35, 145; art museums, 107–8, 110–114, 116, 118, 122, 125n37, 126n43; visual arts, 4, 107, 110–113, 117–118
assemblage, 1, 112, 120, 205, 209, 214

baseline, 5, 162, 165, 170, 176
Black Snake, 52, 57, 59

Calvin, John, 184, 193; Calvinism, 204
capitalism, 12, 30, 38, 41, 47, 64, 67–68, 70, 72–75, 77–78, 149–51, 153, 194, 196, 202, 204–5, 207–11, 213–14; neoliberal capitalism, 2, 4, 6, 48, 66–67, 79; global capitalism, 5, 27, 38, 111, 116, 143, 184–85, 202, 207
capitalocene, 68, 79, 201, 204
centuriation, 129–30, 137
Christianity, 2, 4, 52, 64–66, 68, 74
citizenship, 5–6, 22, 41, 113, 127–30, 134–37
climate change, 1, 9–11, 13–14, 20, 59, 67, 79, 120, 201–2, 204–6, 208, 210, 213. *See also* global warming
colonization, 23, 129–30
commodification, 73, 194–96, 208–9
Communist party, 31–32, 34, 36–38
community of interpreters, 142, 146–47, 149, 155
Congress of Industrial Organizations (CIO), 27, 32–33, 36
cooperativity, 162, 171–77
Corrington, Robert, 5, 142–47, 149, 155

covenant, 5, 162, 167–70
COVID, 2, 111–112, 125n37, 165
credit, 85, 87, 99, 185, 191, 193–94; creditor, 4, 84–85, 87, 91, 96, 101
curation, 4, 108–9, 116, 215; curatorial, 108–9, 113–23

Day Zero, 7, 9, 24
debt, 4, 83–90, 92–101, 102n3, 104n26, 151–52, 186; debt relief, 84, 86; debt forgiveness, 87; debt slavery 84–85, 87–93, 95–96, 99, 102n8
democracy, 1–6, 19, 26–28, 35, 42, 47, 59, 79, 115, 127–28, 131, 142, 145–47, 154, 156, 195, 201, 203, 209–10, 212, 215; economic democracy, 71–73, 77–78
Democratic Socialists of America (DSA), 38, 43–44
depression, 116, 151–52, 165
divine economy, 4, 49, 51, 55, 101, 134–35, 204
DuBois, W. E. B., 78, 206

ecology, 1–2, 5, 65, 67, 73–74, 106–7, 116, 118, 123, 195, 205, 212, 214–15; material ecology, 117, 120–21
economy, 2–4, 26–27, 37, 40–42, 49–51, 54–58, 67, 69, 77–78, 99, 101, 110, 115, 129, 134–36, 140n20, 184–88, 191, 193–96, 203–9
ekklēsia, 128, 131–32, 137, 139n15
entanglements, 2–3, 8–9, 18–20, 22, 24, 136, 183, 188–89, 194–96, 202, 207, 215
Eros, 143, 154–55
eudaimonia, 162, 165. *See also* flourishing

flourishing, 56, 66, 74, 76, 110, 136, 162, 165, 167, 176

gift, 3, 46, 49–58; gift economy, 3, 49–50, 54, 56
global warming, 2, 6, 116, 201–2, 204
Gospel of Prosperity, 4, 64
Grau, Marion, 204
Great Recession, 67, 75

Halberstam, Jack, 5, 143, 148, 150–57
Han, Byeong Chul, 5, 143, 150–51, 154–55

immanence, 4, 65–67, 69–72, 74–76, 78–79
indigenous communities/people, 3, 8, 10, 12–13, 42, 46–50, 52–57, 59, 61n26
Industrial Workers of the World (IWW), 33, 35
interest, 92, 183–186, 190–193, 195–196, 199n47

Jesus (of Nazareth), 6n20, 70, 75, 78, 132, 134, 148, 157, 170, 184, 188, 190, 192–93, 199n49
Jubilee, 84–85, 87, 89, 100, 105n37
Judaism, 66, 87, 139n15; Jewish traditions, 5, 58, 65, 74, 87, 162–64

Keller, Catherine, 5, 6, 9, 17, 106, 109, 113–114, 143, 147–49, 155, 157, 164
King, Martin Luther, Jr., 36, 78
koinonia, 128, 132–33, 136–38, 139n14, 139n15

labor, 4, 6, 15, 27–33, 36, 44, 67–73, 75–79, 84, 86, 92–95, 97, 99, 132, 135, 150, 152–53, 175, 187–88, 196, 209, 214; productive labor, 68, 70; reproductive labor, 68–73, 75–77, 79
Locke, John, 15–16
Luther, Martin, 114, 184, 192–93

Marx, Karl, 30, 35, 67, 70–74, 78, 80n15; Marxism, 28–29, 33, 37–40, 68, 213–14
Matthew, 6, 183, 185, 189–91, 199n49
moneylending, 184, 186–89, 192
Museum of Modern Art (MoMA), 4, 111–112, 116–118, 120, 122, 125n41

natural community, 142, 145–47, 159n24
natural law, 15–17, 21
natural resources, 8, 10–14, 22, 75
New Deal, 28, 32
new materialisms, 4, 65–67, 69, 71, 74

ontology, 5, 25n24, 162–63, 176
Origen, 184, 189–192, 198n28
ownership, 11–12, 14–17, 28, 43, 56, 206; collective ownership, 12
Oxman, Neri, 117–22

Paul, 2, 5, 58, 69, 111, 122, 128–38, 139n14, 139n15
Pally, Marcia, 215
parable, 5–6, 60n4, 183–96, 197n2, 197n8, 199n49, 209
petroleum, 3, 46–47, 51–53, 57; petrocapitalism, 48
Philippi, 129–31, 133, 136–37; Philippians, 5, 128–29, 132–34, 137
politeuma, 5, 128–29, 132–37, 138n6

political theology, 8–9, 16–18, 21, 86, 88, 206, 210
populism, 35
potlatch, 54
production, 4, 48, 57, 65–79, 95–97, 204
property, 3, 7, 9, 11–17, 19, 22, 25n21, 26–27, 47, 56, 84, 100, 131, 134, 171, 189; property rights, 11–12; private property, 7, 9, 11–13, 15–17, 22, 56
pueblos, 3, 6, 9, 21–23

queer art of failure, 5, 143, 154, 157
Quigley, Jennifer, 189

reciprocity, 3, 50–52, 55–59, 69, 168–69
relationality, 5, 50, 94, 149, 162, 164–67, 169–70, 176–77
Revelation, 109, 120, 207; John's *Apocalypse*, 202
Rieger, Joerg, 215
Rome, 88, 129–30, 135, 137, 139n9, 141n26, 187

salvation, 66, 74, 84, 89, 148, 189–90, 205, 210–12, 214
San Bartolo, 9–10, 21, 23
scapegoating, 144–46
Schmitt, Carl, 17, 85, 145, 149, 156, 206
Shachtmanites, 36–38
Singh, Devin, 140n20, 141n21, 185, 189
socialism, 27–36, 41; democratic socialism, 3, 26–28, 32–34, 36, 41, 43–44, 212; Christian socialism, 3, 27–28, 33–36, 81n23; Socialist Party, 27–37
sovereignty, 4, 6, 8, 15, 17, 23, 52–53, 83, 85–86, 88–91, 96, 99–101, 112, 195–96
Standing Rock, 12, 14, 52, 54, 57

Tanner, Kathryn, 4, 6, 195, 202–4, 207, 210–11, 218n32
territorial rights, 3, 8–15, 17–19, 22
theo-economics, 128, 133–35, 137, 188–89, 194, 196, 197n8
transcendence, 4, 67, 69, 71–72, 74–79, 113
transference, 143–44, 146

tribalism, 5, 142–44, 146, 149–50, 152–58
Trinity, 5, 106, 162, 165–67

unions, 27–33, 37, 42
usury, 183, 188, 191–94

water cycle, 3, 8, 10, 18, 20
white nationalism, 26–27

TRANSDISCIPLINARY THEOLOGICAL COLLOQUIA

Laurel Kearns and Catherine Keller, eds., *Ecospirit: Religions and Philosophies for the Earth.*

Virginia Burrus and Catherine Keller, eds., *Toward a Theology of Eros: Transfiguring Passion at the Limits of Discipline.*

Ada María Isasi-Díaz and Eduardo Mendieta, eds., *Decolonizing Epistemologies: Latina/o Theology and Philosophy.*

Stephen D. Moore and Mayra Rivera, eds., *Planetary Loves: Spivak, Postcoloniality, and Theology.*

Chris Boesel and Catherine Keller, eds., *Apophatic Bodies: Negative Theology, Incarnation, and Relationality.*

Chris Boesel and S. Wesley Ariarajah, eds., *Divine Multiplicity: Trinities, Diversities, and the Nature of Relation.*

Stephen D. Moore, ed., *Divinanimality: Animal Theory, Creaturely Theology.* Foreword by Laurel Kearns.

Melanie Johnson-DeBaufre, Catherine Keller, and Elias Ortega-Aponte, eds., *Common Goods: Economy, Ecology, and Political Theology.*

Catherine Keller and Mary-Jane Rubenstein, eds., *Entangled Worlds: Religion, Science, and New Materialisms.*

Kent L. Brintnall, Joseph A. Marchal, and Stephen D. Moore, eds., *Sexual Disorientations: Queer Temporalities, Affects, Theologies.*

Karen Bray and Stephen D. Moore, eds., *Religion, Emotion, Sensation: Affect Theories and Theologies.*

Clayton Crockett and Catherine Keller, eds., *Political Theology on Edge: Ruptures of Justice and Belief in the Anthropocene.*

Kenneth N. Ngwa, Aliou Cissé, and Arthur Pressley, eds., *Life Under the Baobab Tree: Africana Studies and Religion in a Transitional Age.*

Jennifer Quigley and Catherine Keller, eds., *Assembling Futures: Economy, Ecology, Democracy, and Religion.*

www.ingramcontent.com/pod-product-compliance
Lightning Source LLC
Chambersburg PA
CBHW020405080526
44584CB00014B/1184